T0140332

Wireless Networks

Series Editor

Xuemin Sherman Shen, University of Waterloo, Waterloo, ON, Canada

The purpose of Springer's Wireless Networks book series is to establish the state of the art and set the course for future research and development in wireless communication networks. The scope of this series includes not only all aspects of wireless networks (including cellular networks, WiFi, sensor networks, and vehicular networks), but related areas such as cloud computing and big data. The series serves as a central source of references for wireless networks research and development. It aims to publish thorough and cohesive overviews on specific topics in wireless networks, as well as works that are larger in scope than survey articles and that contain more detailed background information. The series also provides coverage of advanced and timely topics worthy of monographs, contributed volumes, textbooks and handbooks.

** Indexing: Wireless Networks is indexed in EBSCO databases and DPLB **

More information about this series at http://www.springer.com/series/14180

Zhenyu Zhou • Zheng Chang • Haijun Liao

Green Internet of Things (IoT): Energy Efficiency Perspective

 Springer

Zhenyu Zhou
North China Electric Power University
Beijing, Beijing, China

Zheng Chang
Faculty of Information Technology
University of Jyväskylä
Jyvaskyla, Finland

Haijun Liao
North China Electric Power University
Beijing, Beijing, China

ISSN 2366-1186   ISSN 2366-1445 (electronic)
Wireless Networks
ISBN 978-3-030-64056-9   ISBN 978-3-030-64054-5 (eBook)
https://doi.org/10.1007/978-3-030-64054-5

This Springer imprint is published by the registered company Springer Nature Switzerland AG
The registered company address is: Gewerbestrasse 11, 6330 Cham, Switzerland

Preface

With the rapid development of information and communication technologies (ICT), current Internet will radically evolve into a ubiquitous network of interconnected objects, which not only collects information from the environment and interacts with the physical world but also provides services for information transfer, analytics, and applications. The recent advances in terms of available computing resources, software systems, and communication networks have made it possible to integrate ICT technologies into virtually anything, thus leading to the emergence of a new paradigm known as the Internet of Things (IoT). The IoT aims to bridge the physical and digital worlds through utilizing advanced sensing and ICT technologies and realize the interconnection of everything. Simultaneously, it offers a platform that enables sensors and devices to connect seamlessly in an intelligent environment, thus providing advanced intelligent services for humans. These intelligence services cover both the communication infrastructure and applications, including monitoring systems, industrial automation, and ultimately smart cities. Until 2019, IoT connections have reached 12 billion globally. With the commercial applications of 5G networks and the ultra-wide coverage of low-power wide-area IoT, the number of connected devices will reach almost 25 billion globally by 2025. This clearly demonstrates the great potential and importance of IoT and is also the motivation behind this book.

However, the huge potential of IoT is constrained by high energy consumption, limited battery capacity, and the slow progress of battery technology. The rapid energy consumption of IoT will cause communication interruption, information loss, and short network lifetime. Moreover, once deployed, the batteries inside IoT devices cannot be replaced in time. Therefore, it is urgent to investigate the related technologies to improve the energy efficiency of IoT, i.e., green IoT technologies. This book seeks to provide a holistic coverage of the green IoT techniques by presenting its principles, enabling technologies, and some typical application domains, which include device-to-device (D2D) communications, machine-to-machine (M2M) communications, space-air-ground networks, and Internet of vehicle (IoV). Some key research issues in green IoT such as networking, resource allocation, and cross-layer resource scheduling are also highlighted. This book also presents some applications discussing the concepts and technologies, which can be leveraged and put to practical usage to solve some real-world problems.

This book is a cohesive material that is comprised of 12 chapters authored by several internationally renowned researchers. Each chapter focuses on a specific subject and also provides the reader with the necessary background information, thus improving understandability and encouraging the reader to think further.

The book may be used as a textbook for both undergraduate and graduate students. It also comes in handy as a reference for researchers, IT professionals, and engineers who are interested in IoT concepts, technologies, and possible applications. We hope the readers will enjoy reading this book as much as we enjoyed reviewing and editing the manuscripts.

Waterloo, ON, Canada   Xuemin Sherman Shen

Contents

1   Introduction 1

2   Energy-Efficient Resource Allocationin for D2D Enabled
    Cellular Networks 5
    2.1   Energy-Efficient Resource Allocation Problem.................... 5
          2.1.1   System Model.. 5
          2.1.2   Problem Formulation 7
    2.2   Energy-Efficient Stable Matching for D2D Communications 10
          2.2.1   Preference Establishment 10
          2.2.2   Energy-Efficient Stable Matching 15
    2.3   Performance Results and Discussions.............................. 16

3   Energy Harvesting Enabled Energy Efficient Cognitive
    Machine-to-Machine Communications 23
    3.1   Framework of Energy-Efficient Resource Allocation for
          EH-Based CM2M 23
          3.1.1   Data Transmission Model 26
          3.1.2   Energy Harvesting and Energy Consumption Model 27
          3.1.3   Energy Efficient Resource Allocation Problem
                  Formulation 27
    3.2   Energy Efficient Joint Channel Selection, Peer Discovery,
          Power Control and Time Allocation for EH-CM2M
          Communications 29
          3.2.1   Matching Based Problem Transformation................. 29
          3.2.2   First-Stage Joint Power Control and Time
                  Allocation Optimization.................................. 30
          3.2.3   Preference List Construction 35
          3.2.4   Second-Stage Joint Channel Selection and Peer
                  Discovery Based on Matching 35
    3.3   Performance Results and Discussions.............................. 37
          3.3.1   Improve Average Energy Efficiency of M2M-TXs 38
          3.3.2   Improve Average Energy Efficiency of M2M Pairs....... 40

4   Software Defined Machine-to-Machine Communication for Smart Energy Management in Power Grids 43
 4.1   Framework of Energy-Efficient SD-M2M for Smart Energy Management 43
  4.1.1   Architecture Overview 43
  4.1.2   The Benefits of the SD-M2M 45
 4.2   Software-Defined M2M Communication for Smart Energy Management Applications 45
 4.3   Case Study and Analysis................................... 48
  4.3.1   Improve Spectral Efficiency............................. 49
  4.3.2   Reduce the Total Energy Generation Cost................ 50

5   Energy-Efficient M2M Communications in for Industrial Automation 53
 5.1   Framework of Energy-Efficient M2M Communications 53
 5.2   Contract-Based Incentive Mechanism Design for Access Control 54
  5.2.1   MTC Type Modeling 54
  5.2.2   Contract Formulation 55
  5.2.3   Contract Optimization 56
 5.3   Resource Allocation Base on Lyapunov Optimization and Matching Theory 58
  5.3.1   Dynamic Queue Model.................................... 58
  5.3.2   Problem Formulation and Transformation 59
  5.3.3   Joint Rate Control, Power Allocation and Channel Selection 61
 5.4   Performance Results and Discussions........................... 64
  5.4.1   Feasibility and Efficiency of Access Control Mechanism... 65
  5.4.2   Feasibility and Efficiency of Resource Allocation Scheme 66

6   Energy-Efficient Context-Aware Resource Allocation for Edge-Computing-Empowered Industrial IoT 69
 6.1   Framework of Energy-Efficient Edge-Computing-Empowered IIoT 69
  6.1.1   System Model... 69
  6.1.2   Problem Formulation 73
 6.2   Learning-Based Context-Aware Channel Selection for the Single-MTD Scenario... 74
  6.2.1   Lyapunov Based Problem Transformation 74
  6.2.2   SEB-GSI Algorithm for the Ideal Case.................... 76
  6.2.3   SEB-UCB Algorithm for the Nonideal Case.............. 78

6.3 Learning-Based Context-Aware Channel Selection for the
Multi-MTD Scenario... 79
6.3.1 SEB-MGSI Algorithm for the Ideal Case 79
6.3.2 SEBC-MUCB Algorithm for the Nonideal Case.......... 82
6.4 Performance Results and Discussions............................. 82
6.4.1 Performance Under the Single-MTD Scenario............ 83
6.4.2 Performance Under the Multi-MTD Scenario............. 85

7 Licensed and Unlicensed Spectrum Management for
Energy-Efficient Cognitive M2M 89
7.1 Framework of CM2M Network 89
7.1.1 System Model... 89
7.1.2 Problem Formulation 93
7.2 Context-Aware Learning-Based Channel Selection for CM2M.... 94
7.2.1 Problem Transformation 94
7.2.2 C^2-GSI for Channel Selection with GSI.................. 95
7.2.3 C^2-EXP3 for Channel Selection with Local
Information 97
7.3 Performance Results and Discussions............................. 99

8 Energy-Efficient Task Assignment and Route Planning for UAV 105
8.1 Framework of UAV-Aided MCS Systems.......................... 105
8.1.1 The Utility Function of the MCS Carrier.................. 107
8.1.2 The Utility Function of UAVs.............................. 108
8.1.3 UAV-Aided MCS Systems Problem Formulation 111
8.2 Energy-Efficient Joint Task Assignment and Route Planning...... 112
8.2.1 Problem Transformation 112
8.2.2 The Route Planning 113
8.2.3 Preference List Construction 117
8.2.4 GS Based Second-Stage Task Assignment 118
8.3 Performance Results and Discussions............................. 120

9 Energy-Efficient and Secure Resource Allocation for
Multiple-Antenna NOMA with Wireless Power Transfer.............. 125
9.1 Framework of Energy-Efficient and Secure Resource
Allocation for Multiple-Antenna NOMA with Wireless
Power Transfer 125
9.1.1 System Model... 125
9.1.2 Problem Formulation 130
9.2 The Energy-Efficient and Secure Resource Allocation Scheme ... 131
9.2.1 Transformation of the Optimization Problem 132
9.2.2 Proposed Algorithmic Solution 133
9.3 Performance Evaluation... 137
9.3.1 Improve Secure Data Rate 137
9.3.2 Improve the Energy Efficiency............................. 140

**10   Dynamic Computation Offloading Scheme for Fog Computing
      System with Energy Harvesting Devices** 143
      10.1   Framework of Socially Aware Dynamic Computation
             Offloading for Fog Computing System with EH Devices 143
             10.1.1   System Movel 143
             10.1.2   Problem Formulation 149
      10.2   Proposed Solution 153
      10.3   Performance Evaluation 156

**11   Energy-Efficient Resource Allocation for Wireless Powered
      Massive MIMO System with Imperfect CSI** 163
      11.1   Framework of Resource Allocation for Wireless Powered
             Massive MIMO System with Imperfect CSI 163
             11.1.1   System Model 163
             11.1.2   Throughput Analysis 165
             11.1.3   Problem Formulation 167
      11.2   Proposed Antenna Selection and Resource Allocation Scheme ... 168
             11.2.1   Proposed Antenna Selection Algorithm 169
             11.2.2   Power and Time Allocation Schemes 169
      11.3   Performance Evaluation 172

12   Summary 177

References 179

List of Acronyms

ICT	Information and communication technologies
IoT	Internet of Things
D2D	Device-to-device
M2M	Machine-to-machine
IoV	Internet of vehicle
BSs	Base stations
QoS	Quality of service
IIoT	Industrial Internet of Things
MTDs	Machine-type devices
3GPP	Third Generation Partnership Project
UAVs	Unmanned aerial vehicles
EE	Energy efficiency
UEs	User equipments
GS	Gale-Shapley
CUs	Cellular users
FDD	Frequency division duplexing
ADC	Analog-to-digital converter
DAC	Digital-to-analog converter
SE	Spectrum efficiency
KKT	Karush-Kuhn-Tucker
CDFs	Cumulative distribution functions
RBs	Resource blocks
P2P	Peer-to-peer
CSI	Channel state information
SINR	Signal to interference plus noise ratio
AO	Alternating optimization
SDN	Software-defined networking
IEDs	Intelligent electronic devices
MDP	Markov decision process
TI	Tactile Internet
MTC	Machine-type communication

ACB   Access class barring
IR   Individual rationality
IC   Incentive compatibility
NOMA   Nonorthogonal multiple access
GSI   Global state information
BER   Bit error rate
BPSK   Binary phase shift keying
QAM   Quadrature amplitude modulation
OFDM   Orthogonal frequency division multiplexing
MAB   Multi-armed bandit
PUs   Primary users
SUs   Secondary users
IQR   Interquartile range
MCS   Mobile crowd sensing
DP   Dynamic programming
GA   Genetic algorithms
AWGN   Additive white Gaussian noise
FIFO   First-in-first-out
MMSE   Minimum mean square error
MRT   Maximum ratio transmission
LPWA   Low power wide area

Chapter 1
Introduction

IoT lays the foundation for the next new industrial revolution that links the Internet with everyday physical objects [1, 2]. It creates a world-wide network of interconnected things that can exchange information amongst and work together with each other to support new applications such as smart city, intelligent transportation system, and smart grid. With the rapid growth of IoT applications, the communications among IoT devices need to be realized anytime and anywhere, which inevitably consume a large portion of battery energy [3]. On the other hand, it is well known that IoT applications often require energy-constrained devices working for long periods without human intervention after their initial deployments. Without efficient resource allocation and energy management mechanisms to save power, these devices would drain their batteries within short time [4]. Therefore, how to reduce the energy consumption and find new ways of deploying green communications for IoT networks has become a critical challenge [5]. Green IoT envisions the concept of reducing the energy consumption of IoT devices and making the IoT networks sustainable [6]. Taking energy efficiency and energy consumption as key design parameters, the implementation of green IoT technologies will reduce the energy consumed for billions of IoT devices communicating with each other, increase resource utilization efficiency and guarantee the QoS requirements of different services. In recent years, green IoT has attracted intensive research interests from both academic and industry [3]. Due to the explosive growth of IoT and 5G cellular technologies, it is predicted that billions of devices will be interconnected and the corresponding data traffic will grow more than 1000 times by 2020 [7–10]. D2D communications that enable ubiquitous information acquisition and exchange among devices over a direct link [11], is a key enabler to facilitate future 5G mobile systems [12]. It not only enables resource allocation under the unified control and scheduling of base stations (BSs), but also supports decentralized data transmission without infrastructures. The D2D devices adopt the technologies of device discovery, mode selection and resource allocation to realize the short-range communications, which brings numerous benefits to

© Springer Nature Switzerland AG 2021
Z. Zhou et al., *Green Internet of Things (IoT): Energy Efficiency Perspective*,
Wireless Networks, https://doi.org/10.1007/978-3-030-64054-5_1

improve network capacity and ensure quality of service (QoS) [13, 27]. Besides, D2D communication can also realize lower energy consumption and shorter delay through spectrum reusing [11]. There already exist numerous research attempts which employ D2D communications to realize green IoT. In [14], Sheng et al. proposed a resource allocation algorithm to maximize the energy efficiency of D2D communication underlaying cellular networks subject to the time-average and network stability constraints by combining fractional programming and Lyapunov optimization. In [15], Peng et al. considered the energy efficiency optimization problem in multimedia heterogeneous cloud radio access networks (HCRANs) subject to individual front-haul capacity as well as multiple interference constraints to sense queue and proposed an online resource allocation algorithm based on Lyapunov optimization. In [16], Babun et al. employed D2D communications to extend the coverage area of active BS, which significantly improves both the energy and spectral efficiency performance compared to conventional cellular networks.

The fourth industrial revolution aims to realize interconnected, responsive, intelligent and self-optimizing manufacturing processes and systems through seamless integration of advanced manufacturing techniques with industrial Internet of Things (IIoT) [17]. In this new paradigm, billions of machine-type devices (MTDs) will be deployed in the field for continuously performing various tasks such as monitoring, billing, and protection [18, 19]. M2M communication technology gives the machine the ability to collect, process and exchange information by installing sensing, information processing and communication devices. Unlike the D2D communications mentioned above, M2M communications focus on the data exchange among machines, which enable devices to communicate autonomously under industrial level QoS constraints [20], and allows the coexistence of human-to-human communications in the same network through establishing a connection between the random-access channel (RACH) and the centralized BSs. The rapidly developing cellular technologies such as 5G have become the major driving forces for the successful implementation of M2M communication due to the ubiquitous presence of cellular infrastructures and the availability of large capacity long-range wide access. Simultaneously, Third Generation Partnership Project (3GPP) has been continuously promoting the international standardization of M2M communications in the current Release-15 and Release-16 specifications [22]. As a key enabler for the successful implementation of green IoT [21], many research efforts are dedicated to address the energy consumption and energy efficiency problems of M2M communications. In [23], a joint time allocation and power control algorithm was proposed for realizing energy-efficient M2M communications with energy harvesting capabilities. In [24], Li et al. introduced MEC into virtualized cellular networks with M2M communications, where each MTC device chooses to access virtual networks so as to minimize the energy consumption and execution time. In [25], Zhang et al. developed a joint power control and time allocation algorithm to minimize the total energy consumption of the M2M network.

Despite the recent progress in conventional D2D and M2M communications, how to realize green communications for IoT, especially industrial IoT with strict delay and reliability requirements remains an open issue. To motivate more researches to devote their efforts into this exciting field, this book aims to provide a comprehensive introduction of fundamental theories and technologies related to green IoT from the perspective of resource allocation and energy management. The organization of this book is as follows. Chapter 1 gives the introduction of green IoT, D2D communications, M2M communications, and the related research progresses. The energy-efficiency optimization problem of D2D communications is addressed in Chap. 2. In Chap. 3, resource allocation optimization for maximizing the energy efficiency of M2M communications is introduced. A software-defined energy-efficient M2M communication framework is presented in Chap. 4. In Chap. 5, a two-stage access control and resource allocation algorithm for energy-efficient M2M communications in industrial automation is investigated. In Chap. 6, an energy-efficient context-aware resource allocation framework for edge-computing-empowered industrial IoT is introduced. The licensed and unlicensed spectrum management problem for energy-efficient cognitive M2M is addressed in Chap. 7. In Chap. 8, the concept of green IoT is extended from conventional ground networks to aerial networks composed of unmanned aerial vehicles (UAVs). An energy-efficient joint route planning and task assignment approach is developed for Internet of UAVs-aided mobile crowd sensing (MCS) systems is demonstrated. In Chap. 9, an energy-efficient and secure resource allocation algorithm design for NOMA systems empowered by wireless power transfer is investigated. In Chap. 10, a socially aware dynamic computation offloading scheme for fog computing system with energy harvesting mobile devices is advocated. In Chap. 11, an energy-efficient resource allocation scheme for a wireless power transfer enabled multi-user massive MIMO system under imperfect channel estimation is proposed. Chapter 12 gives the challenges and future directions of new technologies in terms of energy efficiency, including space-air-ground integrated networks, edge computing, 6G artificial intelligence, and big data.

Chapter 2
Energy-Efficient Resource Allocationin for D2D Enabled Cellular Networks

2.1 Energy-Efficient Resource Allocation Problem

In this section, we firstly provide a detailed description of the system model of D2D communications underlaying cellular networks with UE preferences, and then present the formulation of the energy-efficient resource allocation problem.

2.1.1 System Model

We consider uplink spectrum sharing in D2D communications underlaying cellular networks, which is shown in Fig. 2.1. Uplink spectrum sharing is considered in particularly because firstly, uplink spectrum resources are usually under-utilized compared to the downlink in frequency division duplexing (FDD) based cellular systems [26]; secondly, co-channel interference caused by D2D UEs can be handled more easily by a powerful base station (BS) than CUs. We assume that each CU is allocated with an orthogonal channel (e.g., an orthogonal resource block in LTE), i.e., K active CUs occupy a total of K orthogonal channels and there is no co-channel interference among CUs. A pair of D2D transmitter and receiver that meet D2D communication requirements form a D2D pair, and are allowed to reuse at most one CU's channel for transmission. The scenario that each D2D pair reuses more than one channel is equivalent to a *one-to-many matching* problem, which is out of the scope of this chapter.

As a result, all of active D2D transmitters cause co-channel interference to the BS, and a CU causes co-channel interference to the D2D receiver that operates in the same channel. QoS requirements are imposed for both CUs and D2D pairs. In this chapter, we assume that the D2D mode and peer selection process has already been finished, and we mainly focus on the resource allocation part. The joint optimization of mode selection, peer selection, and resource allocation is a

© Springer Nature Switzerland AG 2021
Z. Zhou et al., *Green Internet of Things (IoT): Energy Efficiency Perspective*,
Wireless Networks, https://doi.org/10.1007/978-3-030-64054-5_2

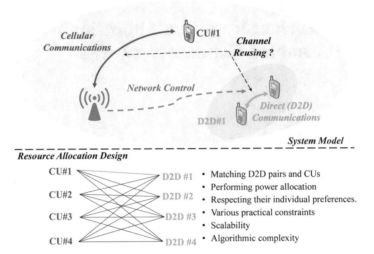

Fig. 2.1 Practical implementation and resource allocation design for D2D communications underlaying cellular networks with UE preferences

completely new problem, which is out of the scope of this chapter. Regarding how to perform mode selection and peer selection, interested readers may refer to [27–29] and references therein for more details.

There are a total of N D2D pairs and K CUs. Throughout the chapter, the index sets of active D2D pairs and CUs are denoted as $\mathcal{D} = \{d_1, \cdots, d_i, \cdots, d_N\}$, and $C = \{c_1, \cdots, c_k, \cdots, c_K\}$, respectively.

Definition 2.1 The partner selection matrix of D2D pairs is denoted as $\mathbf{X}_{N \times K}$, where the (i, k)-th element $x_{i,k} \in \{0, 1\}$ indicates the selection decision of the D2D-CU partnership (d_i, c_k) for the D2D pair d_i, $\forall d_i \in \mathcal{D}$, $\forall c_k \in C$. If $x_{i,k} = 1$, d_i prefers to forming a partnership with c_k, and if $x_{i,k} = 0$, otherwise.

Definition 2.2 The partner selection matrix of CUs is denoted as $\mathbf{Y}_{K \times N}$, where the (k, i)-th element $y_{k,i} \in \{0, 1\}$ indicates the selection decision of the D2D-CU partnership (c_k, d_i) for the CU c_k, $\forall c_k \in C$, $\forall d_i \in \mathcal{D}$. If $y_{k,i} = 1$, c_k prefers to forming a partnership with d_i, and if $y_{k,i} = 0$, otherwise.

Remark 2.1 Due to the individualized and differentiated preferences of d_i and c_k, it is very possible to have conflicting partner selection decisions, i.e., $x_{i,k} \neq y_{k,i}$. A D2D-CU partnership (d_i, c_k) can be formed if and only if $x_{i,k} = y_{k,i} = 1$.

Regarding channel models, both fast fading and slow fading which are caused by multi-path propagation, shadowing, and pathloss are taken into consideration [30]. The channel gain of the interference from c_k to d_i is given by

$$g_{k,i}^c = \varpi \beta_{k,i}^c \zeta_{k,i}^c d_{k,i}^{-\alpha}, \tag{2.1}$$

where ϖ is the pathloss constant, $\beta_{k,i}^c$ is the fast-fading gain with exponential distribution, $\zeta_{k,i}^c$ is the slow-fading gain with log-normal distribution, α is the pathloss exponent, and $d_{k,i}$ is the transmission distance. In a similar way, we can define the channel gain of d_i as g_i^d, the interference channel gain between the transmitter of d_i and the BS as $g_{i,B}^d$, and define the channel gain between c_k and the BS as $g_{k,B}^c$.

The achievable spectrum efficiency (SE) (defined as bits/s/Hz) of d_i is given by

$$U_i^d = \sum_{c_k \in C} \log_2 \left(1 + \frac{x_{i,k} y_{k,i} p_i^d g_i^d}{N_0 + x_{i,k} y_{k,i} p_k^c g_{k,i}^c} \right), \tag{2.2}$$

where p_i^d and p_k^c represent the transmission power of d_i and c_k, respectively. N_0 is the noise power on each channel. The achievable SE of c_k is given by

$$U_k^c = \log_2 \left(1 + \frac{p_k^c g_{k,B}^c}{N_0 + \sum_{d_i \in \mathcal{D}} x_{i,k} y_{k,i} p_i^d g_{i,B}^d} \right). \tag{2.3}$$

The total power consumptions of d_i and c_k are given by

$$E_i^d = \sum_{c_k \in C} \frac{1}{\eta} x_{i,k} y_{k,i} p_i^d + 2p_{cir}, \tag{2.4}$$

$$E_k^c = \frac{1}{\eta} p_k^c + p_{cir}. \tag{2.5}$$

p_{cir} is the total circuit power consumption which includes values of mixer, frequency synthesizer, digital-to-analog converter (DAC)/analog-to-digital converter (ADC), etc. η is the power amplifier (PA) efficiency, i.e., $0 < \eta < 1$. The power consumption of the BS is not considered because it is powered by external grid power.

2.1.2 Problem Formulation

Eventually better channel conditions and proper transmission power strategies can improve the EE performance more efficiently. Therefore, for any D2D pair or CU, the following questions need to be answered before reaching a decision:

- How to select a partner to form a D2D-CU channel-reusing pair for optimizing EE performance?
- How to perform power allocation for the expected D2D-CU pair?
- How to satisfy various practical resource allocation constraints such as maximum transmission power levels, QoS requirements, and channel-reusing rules, etc?

- How to avoid disruptions from other D2D pairs or CUs which also wish to be matched with the preferred D2D pair or CU?

The above questions indicate that the optimization of EE involves solving a *joint partner selection and power allocation problem*. To be more general, denoting $\mathbf{x_i} = \{x_{i,1}, \cdots, x_{i,k}, \cdots, x_{i,K}\}$ and $\mathbf{y_k} = \{y_{k,1}, \cdots, y_{k,i}, \cdots, y_{k,N}\}$ as d_i's and c_k's binary partner selection strategy sets, respectively, and denoting p_i^d and p_k^c as d_i's and c_k's continuous power allocation strategies, respectively, the objective function in terms of EE (bits/J/Hz) is defined as the spectrum efficiency (SE) (bits/s/Hz) divided by the total power consumption (W) [31]. The EE objective functions of d_i (including both the transmitter and receiver) and c_k are given by

$$U_{i,EE}^d(\mathbf{x_i}, p_i^d) = \frac{U_i^d(\mathbf{x_i}, p_i^d)}{E_i^d(\mathbf{x_i}, p_i^d)}$$

$$= \frac{\sum_{c_k \in C} \log_2 \left(1 + \frac{x_{i,k} y_{k,i} p_i^d g_i^d}{N_0 + x_{i,k} y_{k,i} p_k^c g_{k,i}^c}\right)}{\sum_{c_k \in C} \frac{1}{\eta} x_{i,k} y_{k,i} p_i^d + 2p_{cir}}, \tag{2.6}$$

$$U_{k,EE}^c(\mathbf{y_k}, p_k^c) = \frac{U_k^c(\mathbf{y_k}, p_k^c)}{E_k^c(p_k^c)}$$

$$= \frac{\log_2 \left(1 + \frac{p_k^c g_{k,B}^c}{N_0 + \sum_{d_i \in D} x_{i,k} y_{k,i} p_i^d g_{i,B}^d}\right)}{\frac{1}{\eta} p_k^c + p_{cir}}. \tag{2.7}$$

The joint partner selection and power allocation problem for d_i can be formulated as

$$\begin{aligned}
&\max_{(\mathbf{x_i}, p_i^d)} && U_{i,EE}^d(\mathbf{x_i}, p_i^d) \\
&\text{s.t.} && C_{i,1}^d : 0 \le p_i^d \le p_{i,max}^d, \\
& && C_{i,2}^d : U_i^d(\mathbf{x_i}, p_i^d) \ge U_{i,min}^d, \\
& && C_{i,3}^d : x_{i,k} = \{0, 1\}, \forall c_k \in C, \\
& && C_{i,4}^d : \sum_{c_k \in C} x_{i,k} \le 1. \tag{2.8}
\end{aligned}$$

$C_{i,1}^d$ ensures that the power allocation of d_i should not exceed the maximum allowed transmission power $p_{i,max}^d$. $C_{i,2}^d$ specifies the QoS requirement which represents that the minimum SE should not fall below $U_{i,min}^d$. $C_{i,3}^d$ and $C_{i,4}^d$ are the channel-reusing constraints which make sure that at most one channel can be shared simultaneously by d_i and one existing CU.

The problem formulation for c_k is given by

$$\max_{(\mathbf{y_k}, p_k^c)} \quad U_{k,EE}^c(\mathbf{y_k}, p_k^c)$$

$$\text{s.t.} \quad C_{k,1}^c : 0 \leq p_k^c \leq p_{k,max}^c,$$

$$C_{k,2}^c : U_k^c(\mathbf{y_k}, p_k^c) \geq U_{k,min}^c,$$

$$C_{k,3}^c : y_{k,i} = \{0, 1\}, \forall d_i \in \mathcal{D},$$

$$C_{k,4}^c : \sum_{d_i \in \mathcal{D}} y_{k,i} \leq 1. \tag{2.9}$$

$C_{k,1}^c$ and $C_{k,2}^c$ specify the transmission power and QoS constraints. $C_{k,3}^c$ and $C_{k,4}^c$ ensure that at most one D2D pair can share the same channel with c_k simultaneously.

Remark 2.2 By observing (2.8) and (2.9), we find that the partner selection problem is coupled with the power allocation problem. The formulation obtained is an NP-hard mixed integer nonlinear programming (MINLP) problem, which involves both binary and continuous variables for resource allocation optimization. Thus, the formulations obtained in neither (2.8) nor (2.9) cannot be solved directly by using either nonlinear fractional programming or integer programming. Furthermore, neither of the two programming approaches has taken UEs' preferences and satisfactions into consideration, which may lead to unstable and unsatisfied resource allocation decision.

To solve (2.8) and (2.9), we introduce a *one-to-one matching* model to match D2D pairs with CUs according to their mutual preferences. In this fashion, the original NP-hard MINLP problem can be decoupled into two separate subproblems and solved in a tractable manner. We use the triple $(C, \mathcal{D}, \mathcal{P})$ to denote the formulated matching problem, i.e., $\mathcal{D}$, C represent the two finite and distinct sets of D2D pairs and CUs, respectively, and $\mathcal{P}$ is the set of mutual preferences. Both D2D pairs and CUs seek to form proper channel-reusing partnerships to maximize EE under constraints of QoS and transmission power. The definition of a matching μ is given by [32]:

Definition 2.3 For the matching problem $(C, \mathcal{D}, \mathcal{P})$, μ is a point-by-point mapping from $C \cup \mathcal{D}$ onto itself under preference $\mathcal{P}$. This is, for any $c_k \in C$ and $d_i \in \mathcal{D}$, $\mu(c_k) \in \mathcal{D} \cup \{c_k\}$ and $\mu(d_i) \in C \cup \{d_k\}$. $\mu(c_k) = d_i$ if and only if $\mu(d_i) = c_k$.

If $\mu(d_i) = d_i$ or $\mu(c_k) = c_k$, d_i or c_k stays single. Either d_i or c_k can send a request for forming a partnership with its preferred partner based on its preference (which is the partner selection subproblem), and demonstrate the allocated transmission power for the formed partnership (which is the power allocation subproblem). Both d_i and c_k are assumed to only care about their own matched partners and show little concerns to matching results of others. This assumption is valid because UEs are privately owned and operated by independent individuals.

2.2   Energy-Efficient Stable Matching for D2D Communications

In this section, we introduce the energy-efficient stable matching approach. First, we develop an iterative algorithm for preference establishment based on noncooperative game theory and nonlinear fractional programming in Sect. 2.2.1. Then, the derivation of the energy-efficient matching based on the GS algorithm is presented in Sect. 2.2.2.

2.2.1   Preference Establishment

(1) Noncooperative Game Based Preference Modeling
The set of UEs' preferences $\mathcal{P}$ is necessary for developing the energy-efficient matching. We model d_i's preference over c_k as the maximum achievable EE under the matching $\mu(d_i) = c_k$ ($x_{i,k} = y_{k,i} = 1$). Thus, the partner selection decision of d_i has been already fixed, and only the power allocation strategy needs to be optimized. The formulated power allocation problem is given by

$$\max_{p_i^d} \quad U_{i,EE}^d(p_i^d)\Big|_{\mu(d_i)=c_k}$$
$$\text{s.t.} \quad C_{i,1}^d, C_{i,2}^d. \tag{2.10}$$

The power allocation problem for c_k under the matching $\mu(c_k) = d_i$ is given by

$$\max_{p_k^c} \quad U_{k,EE}^c(p_k^c)\Big|_{\mu(c_k)=d_i}$$
$$\text{s.t.} \quad C_{k,1}^c, C_{k,2}^c. \tag{2.11}$$

There are two challenges when solving the above optimization problems. First, from (2.6) and (2.7), $U_{i,EE}^d$ and $U_{k,EE}^c$ are inter-correlated through the interference terms, i.e., $p_k^c g_{k,i}^c$ and $p_i^d g_{i,B}^d$. Second, the problems formulated in (2.10) and (2.11) are still nonconvex due to the fractional form of $U_{i,EE}^d$ and $U_{k,EE}^c$.

In order to study the inter-connections between D2D pairs and CUs (to solve the first challenge), we adopt a game-theoretic approach to model the distributed power allocation problem as a noncooperative game $\mathcal{G}$. UEs are assumed as rational and selfish [33], i.e., each $d_i \in \mathcal{D}$ (or $c_k \in C$) cares about its individual objective

$U_{i,EE}^d$ (or $U_{k,EE}^c$), but is not otherwise concerned with $U_{j,EE}^d$, $\forall d_j \in \mathcal{D}\backslash\{d_i\}$ (or $U_{m,EE}^c$, $\forall c_m \in C\backslash\{c_k\}$). The game $\mathcal{G}$ can be described as $\mathcal{G} = (C, \mathcal{D}, \mathcal{A}, \mathcal{U})$, wherein $\mathcal{A} = \{A_1^d, \cdots, A_N^d, A_1^c, \cdots, A_K^d\}$ is the set of possible strategies that a UE can take, and $\mathcal{U} = \{U_{1,EE}^d, \cdots, U_{N,EE}^d, U_{1,EE}^c, \cdots, U_{K,EE}^c\}$ is the set of UEs' utilities. For example, if $A_i^d = \{[0, p_{i,max}^d]\}$, then d_i is allowed to select p_i^d from the interval $[0, p_{i,max}^d]$.

(2) Objective Function Transformation
To overcome the second challenge, nonlinear fractional programming is employed to transform the nonconvex problem in the fractional form to equivalent convex ones. The optimum result of (2.10) is defined as

$$q_i^{d*} = \max_{p_i^d} U_{i,EE}^d(p_i^d)\Big|_{\mu(d_i)=c_k} = \frac{U_i^d(p_i^{d*})}{E_i^d(p_i^{d*})}, \tag{2.12}$$

where p_i^{d*} is the optimum power allocation strategy of d_i. Based on [34], we have

Theorem 2.1 q_i^{d*} *is achieved if and only if*

$$\max_{p_i^d} U_i^d(p_i^d) - q_i^{d*} E_i^d(p_i^d) = U_i^d(p_i^{d*}) - q_i^{d*} E_i^d(p_i^{d*}) = 0. \tag{2.13}$$

Theorem 2.1 reveals that there exists an equivalent transformed problem with an objective function in subtractive form, which leads to the same maximum EE obtained by directly solving (2.10). The equivalent optimization problem in subtractive form is given by

$$\max_{p_i^d} U_i^d(p_i^d) - q_i^{d*} E_i^d(p_i^d)$$

$$\text{s.t.} \qquad C_{i,1}^d, C_{i,2}^d. \tag{2.14}$$

(2.14) is actually a multi-objective convex optimization problem where the variable q_i^{d*} can be regarded as a negative weight of E_i^d. In the same way, defining q_k^{c*} and p_k^{c*} as the optimum EE and the corresponding strategy of c_k, respectively, the transformed problem that is equivalent to (2.11) is given by

$$\max_{p_k^c} U_k^c(p_k^c) - q_k^{c*} E_k^c(p_k^c)$$

$$\text{s.t.} \qquad C_{k,1}^c, C_{k,2}^c. \tag{2.15}$$

(3) Distributed Iterative Power Allocation
Both (2.14) and (2.15) are standard convex optimization problems and can be solved efficiently. However, the specific values of q_i^{d*} and q_k^{c*} are required to solve (2.14) and (2.15), respectively. In order to obtain q_i^{d*} and q_k^{c*}, an iterative algorithm is

developed based on Dinkelbach's method and is given in Algorithm 2.1 [34]. The iterative Algorithm 2.1 consists of two loops: the outer loop with the iteration index l represents iterations of the noncooperative game, and the inner loop with the iteration index n represents iterations of Dinkelbach's algorithm. The relationship between inner loop and outer loop iterations is shown in Fig. 2.2. For each round of the game, the inner loop is executed to find the corresponding optimum power allocation strategy for each player, which stops if either the iteration stopping criteria or the maximum loop number N_{max} is reached. The game iteration continues until the achieved power allocation strategy converges to a Nash equilibrium, i.e., none player is capable of unilaterally achieving better performance by deviating from it.

At the n-th iteration of the l-th round game, $p_i^d(n)$ and $p_k^c(n)$ are obtained by solving the following problems with $q_i^d(n)$ and $q_k^c(n)$ obtained from the $(n-1)$-th iteration:

$$\max_{p_i^d} U_i^d[p_i^d(n)] - q_i^d(n) E_i^d[p_i^d(n)]$$

s.t. $\qquad\qquad C_{i,1}^d, C_{i,2}^d.$ \hfill (2.16)

$$\max_{p_k^c} U_k^c[p_k^c(n)] - q_k^c(n) E_k^c[p_k^c(n)]$$

s.t. $\qquad\qquad C_{k,1}^c, C_{k,2}^c.$ \hfill (2.17)

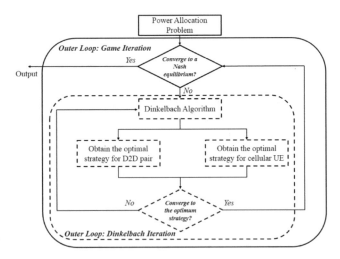

Fig. 2.2 The relationship between inner loop and outer loop iterations of the iterative power allocation algorithm

Algorithm 2.1 Distributed iterative power allocation algorithm for obtaining q_i^{d*} and q_k^{c*}

1: **Input:** g_i^d, $\hat{p}_i^d g_{i,B}^d$, $g_{k,B}^c$, $\hat{p}_k^c g_{k,i}^c$, $p_{i,max}^d$, $p_{k,max}^d$, $C_{i,min}^d$, $C_{k,min}^c$.
2: **Output:** q_i^{d*}, q_k^{c*}, p_i^{d*}, p_k^{c*}.
3: **Initialize:** q_i^d, q_k^c, N_{max}, Δ_g, Δ_d, Δ_c, $\hat{p}_k^c$, $\hat{p}_i^d$
4: **while** $| \mathbf{q}_i^{d*}(l) - \mathbf{q}_i^{d*}(l-1) | \leq \Delta_g$
     & $| \mathbf{q}_k^{c*}(l) - \mathbf{q}_k^{c*}(l-1) | \leq \Delta_g$ **do**
5:     **while** $n < N_{max}$ **do**
6:         obtain $\hat{p}_i^d(n)$ using (2.20)
7:         **if** $U_i^d[\hat{p}_i^d(n)] - q_i^d(n) E_i^d[\hat{p}_i^d(n)] > \Delta_d$ **then**
8:             **Update:** $q_i^d(n+1) = U_i^d[\hat{p}_i^d(n)]/E_i^d[\hat{p}_i^d(n)]$
9:         **else**
10:            $p_i^{d*} = \hat{p}_i^d(n)$, and $q_i^{d*} = U_i^d(p_i^{d*})/E_i^d(p_i^{d*})$
11:        **end if**
12:        obtain $\hat{p}_k^c(n)$ using (2.23)
13:        **if** $U_k^c[\hat{p}_k^c(n)] - q_k^c(n) E_k^c[\hat{p}_k^c(n)] > \Delta_c$ **then**
14:            **Update:** $q_k^c(n+1) = U_k^c[\hat{p}_k^c(n)]/E_k^c[\hat{p}_k^c(n)]$
15:        **else**
16:            $p_k^{c*} = \hat{p}_k^c(n)$, and $q_k^{c*} = U_k^c(p_k^{c*})/E_k^c(p_k^{c*})$
17:        **end if**
18:        **Update the Dinkelbach iteration index:** $n \to n+1$
19:    **end while**
20:    **Update:** $\mathbf{q}_i^d(l) = q_i^{d*}$, $\mathbf{q}_k^{c*}(l) = q_k^{c*}$
21:    **Update the game iteration index :** $l \to l+1$
22: **end while**

The augmented Lagrangian of (2.16) is given by

$$\mathcal{L}_i^{EE}(p_i^d, \delta_i^d, \theta_i^d) = U_i^d[p_i^d(n)] - q_i^d(n) E_i^d[p_i^d(n)]$$
$$- \delta_i^d(n)[p_i^d(n) - p_{i,max}^d] + \theta_i^d(n)\left(U_i^d[p_i^d(n)] - U_{i,min}^d\right), \tag{2.18}$$

where δ_i^d and θ_i^d are the Lagrange multipliers for constraints $C_{i,1}^d$ and $C_{i,2}^d$, respectively. By using Lagrange dual decomposition, (2.18) is decomposed as [35]

$$\min_{(\delta_i^d, \theta_i^d \geq 0)} \quad \max_{(p_i^d)} \quad \mathcal{L}_i^{EE}(p_i^d, \delta_i^d, \theta_i^d). \tag{2.19}$$

By exploiting Karush–Kuhn–Tucker (KKT) conditions, the optimal value $\hat{p}_i^d(n)$ corresponding to $q_i^d(n)$ is given by

$$\hat{p}_i^d(n) = \left[\frac{\eta[1 + \theta_i^d(n)] \log_2 e}{q_i^d(n) + \eta \delta_i^d(n)} - \frac{\hat{p}_k^c g_{k,i}^c(n) + N_0}{g_i^d} \right]^+, \tag{2.20}$$

Algorithm 2.2 Iterative preference establishment algorithm for obtaining $\mathcal{P}$

1: **Input:** the set of CUs and D2D pairs, $C, \mathcal{D}$.
2: **Output:** the set of preference profiles $\mathcal{P}$.
3: **for** $d_i \in \mathcal{D}$ **do**
4:     **for** $c_k \in C$ **do**
5:         calculate maximum achievable $q_i^{d*}\big|_{\mu(d_i)=c_k}$ and $q_k^{c*}\big|_{\mu(c_k)=d_i}$ for the D2D-CU pair
        (d_i, c_k) by employing Algorithm 2.1.
6:     **end for**
7: **end for**
8: **for** $d_i \in \mathcal{D}$ **do**
9:     sort all of CUs $c_k \in C$ in a descending order according to $q_i^{d*}\big|_{\mu(d_i)=c_k}$.
10: **end for**
11: **for** $c_k \in C$ **do**
12:     sort all of D2D pairs $d_i \in \mathcal{D}$ in a descending order according to $q_k^{c*}\big|_{\mu(c_k)=d_i}$.
13: **end for**

---

where $[x]^+ = \max\{0, x\}$. Then, by employing the gradient method [36], we update the Lagrange multipliers as

$$\delta_i^d(n, \tau + 1) = \left[\delta_i^d(n, \tau) + \epsilon_{i,\delta}(n, \tau)\left(\hat{p}_i^d(n, \tau) - p_{i,max}^d\right)\right]^+, \tag{2.21}$$

$$\theta_i^d(n, \tau + 1) = \left[\theta_i^d(n, \tau) - \epsilon_{i,\theta}(n, \tau)\left(U_i^d(n, \tau) - U_{i,min}^d\right)\right]^+, \tag{2.22}$$

where τ is the iteration index of Lagrange multiplier updating, $\epsilon_{i,\delta}$ and $\epsilon_{i,\theta}$ are the step sizes. The step sizes should be carefully chosen to guarantee convergence and optimality.

Then, $\hat{p}_i^d(n)$ obtained in (2.20) is used to update $q_i^d(n + 1)$ for the $(n + 1)$-th iteration as $q_i^d(n + 1) = U_i^d[\hat{p}_i^d(n)]/E_i^d[\hat{p}_i^d(n)]$. In the final iteration of the inner loop, setting $p_i^{d*} = \hat{p}_i^d$, q_i^{d*} can be obtained by using (2.12) and saved as the l-th element of the vector $\mathbf{q}_i^d$, i.e., $\mathbf{q}_i^d(l) = q_i^{d*}$. The optimization problem (2.17) is solved in the same way. The optimal value $\hat{p}_k^c(n)$ corresponds to $q_k^c(n)$ is given by

$$\hat{p}_k^c(n) = \left[\frac{\eta[1 + \xi_k^c(n)]\log_2 e}{q_k^c(n) + \eta\rho_k^c(n)} - \frac{\hat{p}_i^d(n)g_{i,B}^d + N_0}{g_{k,B}^c}\right]^+. \tag{2.23}$$

Details for how to obtain q_k^{c*} are omitted due to space restrictions. The outer loop stops if the maximum EE (q_i^{d*}, q_k^{c*}) obtained in the l-th round of the game varies little from the optimization result achieved in the previous round, where the corresponding optimum strategy set (p_i^{d*}, p_k^{c*}) has converged to a Nash equilibrium.

Algorithm 2.2 presents how to establish the set of preference profiles $\mathcal{P}$. For every $d_i \in \mathcal{D}$, the maximum achievable q_i^{d*} under the matching $\mu(d_i) = c_k, \forall c_k \in C$, is denoted as $q_i^{d*}\Big|_{\mu(d_i)=c_k}$, and can be obtained by using Algorithm 2.1. We write $c_k \succ_{d_i} c_m$ to mean d_i prefers c_k to c_m, which is defined as

$$c_k \succ_{d_i} c_m \Leftrightarrow q_i^{d*}\Big|_{\mu(d_i)=c_k} > q_i^{d*}\Big|_{\mu(d_i)=c_m}, \tag{2.24}$$

where $\succ$ is a complete, reflexive, and transitive binary preference relation [32]. In addition, we write $c_k \succeq_{d_i} c_m$ to mean d_i likes c_k at least as well as c_m, which is defined as

$$c_k \succeq_{d_i} c_m \Leftrightarrow q_i^{d*}\Big|_{\mu(d_i)=c_k} \geq q_i^{d*}\Big|_{\mu(d_i)=c_m}. \tag{2.25}$$

Similarly, we write $d_i \succ_{c_k} d_j$ to mean c_k prefers d_i to d_j, which is defined as

$$d_i \succ_{c_k} d_j \Leftrightarrow q_k^{c*}\Big|_{\mu(c_k)=d_i} > q_k^{c*}\Big|_{\mu(c_k)=d_j}. \tag{2.26}$$

After obtaining $q_i^{d*}\Big|_{\mu(d_i)=c_k}$, $\forall c_k \in C$, the preference profile $P(d_i) = \{\cdots, c_k, c_m \cdots\}$ is obtained by sorting all of CUs in a descending order according to the criteria of $q_i^{d*}\Big|_{\mu(d_i)=c_k}$, $\forall c_k \in C$. The preference profile of c_k is denoted as $P(c_k)$, which is obtained by sorting all of available D2D pairs according to $q_k^{c*}\Big|_{\mu(c_k)=d_i}$, $\forall d_i \in D$. The total set $\mathcal{P}$ is constructed as $\mathcal{P} = \{P(d_1), \cdots, P(d_N), P(c_1), \cdots, P(c_K)\}$.

2.2.2   Energy-Efficient Stable Matching

After obtaining $P(d_i)$ and $P(c_k)$ for each $d_i \in D$ and $c_k \in C$, we introduce Algorithm 2.3 to match D2D pairs with CUs by employing the GS algorithm [37]. In the first iteration, every $d_i \in D$ sends a partner request to its most preferred CU $\max\{q_i^{d*}\Big|_{\mu(d_i)=c_k}$, $\forall c_k \in C\}$. Then, every $c_k \in C$ receives the request and rejects the D2D pair if it already holds a better candidate. Any $d_i \in D$ that is not rejected by the CUs at this step is held as a candidate. In the next step, any $d_i \in D$ that has been already rejected sends a new request to its most preferred choice from the set of CUs that have not yet issued a rejection. If a D2D pair is rejected by all of its preferred CUs, it will give up and send no further request. Each $c_k \in C$ compares all of the received requests including the candidate that was held from previous steps and only accepts the most preferred D2D pair. The request sending and rejection process finishes when every $d_i \in D$ has already found a partner or has been rejected by all of CUs to which it has sent requests. Algorithm 2.3 has the property of *deferred*

Algorithm 2.3 Energy-efficient stable matching algorithm

1: **Input:** $C, \mathcal{D}, \mathcal{P}$.
2: **Output:** μ.
3: **Initialize:** $\mu = \phi$, $\Phi = \mathcal{D}$.
4: **while** $\Phi \neq \phi$ **do**
5:     **for** $d_i \in \Phi$ **do**
6:         d_i chooses the CU with the highest ranking from $P(d_i)$.
7:     **end for**
8:     **for** $c_k \in C$ **do**
9:         **if** c_k receives a request from d_i, and prefers d_i to its current candidate d_j held from previous steps, i.e., $d_i \succ_{c_k} d_j$ **then**
10:             d_i is held as a new candidate, while c_k issues a rejection to d_j, i.e., $\mu(c_k) = d_i$;
11:             add d_j into Φ, remove d_i from Φ, and remove c_k from $P(d_j)$.
12:         **else**
13:             c_k issues a rejection to d_i, and holds d_j continually as its candidate, i.e., $\mu(c_k) = d_j$.
14:             remove c_k from $P(d_i)$.
15:         **end if**
16:     **end for**
17: **end while**

---

acceptance due to the fact that the best candidate kept at any step can be rejected later on if a better candidate appears.

2.3   Performance Results and Discussions

In this section, the energy-efficient matching algorithm, labeled as "energy-efficient stable matchin", is compared with several heuristic algorithms. The first is the power greedy algorithm, which always allocates the maximum transmission power $p_{i,max}^d$ (or $p_{k,max}^c$). The second is the random power allocation algorithm, which allocates power uniformly distributed in the range $[0, p_{i,max}^d]$ (or $[0, p_{k,max}^c]$). The third is the spectrum-efficient algorithm based on the water-filling power allocation (SINR maximization) [38–40]. The first two heuristic algorithms employ random matching which matches D2D pairs with CUs in a random way, while the third one adopts maximum-SINR based association. The values of simulation parameters are based on [30, 38, 41], and are summarized in Table 2.1. A single cellular network is considered and the cell radius is 500 m. In each time of simulation, locations of CUs and D2D pairs are generated in a random way as shown in Fig. 2.3. QoS requirements in terms of SE are generated randomly from a uniform distribution in the range [0.5, 1] bit/s/Hz. We average the simulation results over 10^3 times.

The average EE performance and UE satisfactions of the obtained energy-efficient stable matching is evaluated and verified. We adopt a statistical model to define a UE's satisfaction as the cumulative distribution functions (CDFs) of the matching result that is higher than its satisfaction threshold. For example, defining d_i's satisfaction threshold as c_m, the matching result $\mu(d_i)$ is compared with the

Table 2.1 Simulation parameters

Simulation parameter	Value
Cell radius	500 m
Max D2D transmission distance d_{max}^d	$20 \sim 100$ m
Pathloss exponent α	4
Pathloss constant ϖ	10^{-2}
Shadowing $\zeta_{k,i}^c$ (standard deviation of a log-normal distribution)	8 dB
Multi-path fading $\beta_{i,k}^c$ (the mean of an exponential distribution)	1
Max Tx power $p_{i,max}^d$, $p_{k,max}^c$	23 dBm
Constant circuit power p_{cir}	20 dBm
Noise power N_0	-114 dBm
Number of D2D pairs N	$5 \sim 50$
Number of cellular UEs K	$5 \sim 50$
PA efficiency η	35%
QoS requirement $C_{k,min}^c$, $C_{i,min}^d$ (uniform distribution)	$0.5 \sim 1$ bit/s/Hz

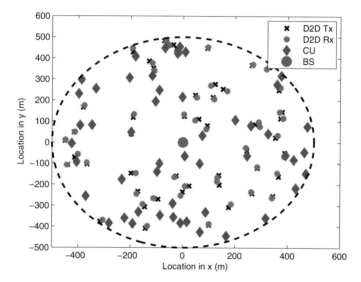

Fig. 2.3 A snapshot of locations of K CUs and N D2D pairs in a single cellular network ($K = N = 50$, the cell radius is 500 m)

threshold c_m to evaluate whether d_i is satisfied with $\mu(d_i)$. d_i is said to be satisfied with $\mu(d_i)$ if d_i prefers $\mu(d_i)$ at least as well as c_m, i.e., $\mu(d_i) \succeq_{d_i} c_m$. Otherwise, d_i is said to be unsatisfied with $\mu(d_i)$ if it is matched to a partner that is less preferred to the threshold, i.e., $c_m \succ_{d_i} \mu(d_i)$. The CDF is denoted as $Pr\{\mu(d_i) \succeq_{d_i} c_m\}$, which is the probability that d_i is matched with a partner that is more preferred to the threshold c_m.

Figure 2.4 shows the average EE performance of D2D pairs versus the maximum D2D transmission distance d_{max}^d with $K = 5$ CUs and $N = 5$ D2D pairs. Simulation

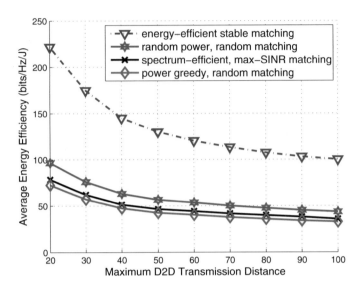

Fig. 2.4 Average EE of D2D pairs versus maximum D2D transmission distance ($K = N = 5$, $d_{max}^d = 20 \sim 100$ m)

results demonstrate that the energy-efficient matching algorithm achieves the best EE performance in the whole regime and outperforms the random power allocation algorithm, the power greedy algorithm, and the spectrum-efficient algorithm by 132, 206, and 248% for $d_{max}^d = 20$ m, respectively. Random allocation achieves the second best performance since there is a large probability to have a higher EE than the spectrum-efficient and power greedy algorithms which always take full advantage of any available power. It is clear that the SE gain achieved by increasing transmission power is not able to compensate for the corresponding EE loss. The power greedy algorithm has the worst EE performance among the four due to two reasons. First, power consumption is completely ignored in the resource allocation process. Second, increasing transmission power beyond the point for optimum SE not only brings no SE improvement in an interference-limited environment but also causes significant EE loss. Note that, as the D2D transmission distance increases, the EE performance of all algorithms decreases because higher transmission power is required to maintain the same QoS performance than in the scenario of short distance.

Figure 2.5 shows the average EE performance of D2D pairs versus the number of active CUs K and D2D pairs N with $d_{max}^d = 20$ m. The average EE performance of all algorithms increases linearly as the active number of CUs and D2D pairs increases. The reason is that as the number of CUs increases, not only the total number of available orthogonal channels increases, but also each D2D pair has a wider variety of choice in the expanded matching market than in the original one. The probability for a D2D pair to be matched with a better partner becomes higher in the expanded matching market. The energy-efficient matching algorithm has the

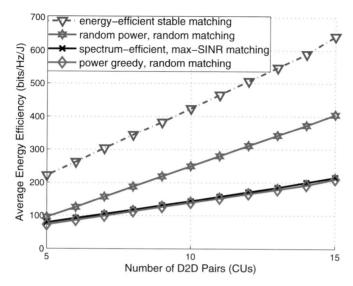

Fig. 2.5 Average EE of D2D pairs versus numbers of active D2D pairs and CUs ($d_{max}^d = 20$ m, $K = N = 5 \sim 15$)

steepest slope among the four, which indicates that it can exploit more benefits from the diversity of choices than the heuristic algorithms could. Both the spectrum-efficient and the power greedy algorithms have the flattest slope since the value of choice diversity is not fully exploited and power consumption is also ignored in the resource allocation process.

Figure 2.6 shows the CDF of D2D pairs' satisfactions versus various satisfaction thresholds with $K = N = 20, 50$ and $d_{max}^d = 20$ m. We adopt the Monte-Carlo method to calculate the CDF that uses repeated matching results (10^4 times) to obtain the numerical results. In the case of $K = N = 20$, the probability of being matched to the first three choices for D2D pairs is 66.4%. In contrast, the corresponding probability under random matching is only 15.4%. When the number of D2D pairs and CUs is increased from 20 to 50, there is still as high as 56.9% of D2D pairs that have been matched to the first three choices, while the corresponding probability under random matching is decreased dramatically from 15.4% to only 6.4%. Significant UE satisfaction gains can be achieved by the energy-efficient matching algorithm compared to the random matching. In addition, the simulation results also reveal the fact that the energy-efficient matching algorithm is able to outperform the random matching for a wide range of satisfaction values.

Figure 2.7 shows the convergence of the iterative algorithm (Algorithm 2.1) versus the number of game iterations. It is shown that the energy-efficient matching algorithm only requires 3 ~ 4 iterations to converge to the equilibrium. In the first game iteration, higher EE performance can be achieved because CUs are not aware of the D2D pairs and transmit in lower power levels. With the entry of D2D pairs into the game, CUs have to increase their transmission power to satisfy QoS

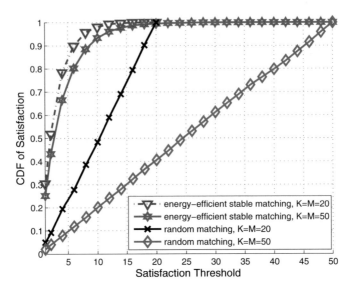

Fig. 2.6 CDF of D2D pairs' satisfactions versus satisfaction threshold ($N = K = 20, 50, d^d_{max} = 20$ m)

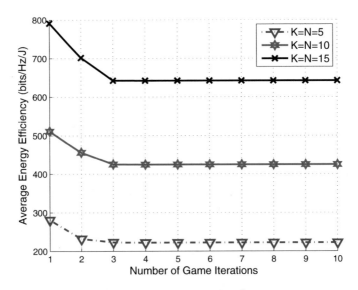

Fig. 2.7 Average EE versus the number of game iterations ($d^d_{max} = 20$ m, $K = N = 5, 10, 15$)

requirements, which in turn causes co-channel interference to D2D pairs and reduce their achievable EE performance until they reach a Nash equilibrium.

In this chapter, an energy-efficient stable matching algorithm was introduced for the resource allocation problem in D2D communications. Taking into account UEs' preferences and satisfactions, the joint partner selection and power allocation problem was formulated to maximize the achievable EE under maximum transmission power and QoS constraints. The resulting problem is nonconvex and computationally intractable. Inspired by the matching theory and game theory, the NP-hard problem was transformed into a one-to-one matching problem with UEs' preferences modeled as the optimum EE under a specific matching. A noncooperative game based iterative algorithm was introduced to establish mutual preferences by exploiting nonlinear fractional programming. Extensive simulation results validate the effectiveness and superiority of the algorithm. The present matching approach sheds new light on the research directions for resource allocation problems in green D2D communications.

Chapter 3
Energy Harvesting Enabled Energy Efficient Cognitive Machine-to-Machine Communications

3.1 Framework of Energy-Efficient Resource Allocation for EH-Based CM2M

Figure 3.1 shows the scenario of EH-CM2M communication underlaying a single-cell cellular network. A block fading channel model [42] is adopted, where the channel state is assumed to be fixed over a time block, and independently changes from one time block to another. In each time block, there are a set of K CUs and a set of N M2M-TXs, which are denoted by $C_K = \{C_1, \ldots, C_k, \ldots, C_K\}$ and $\mathcal{T}X_N = \{TX_1, \ldots, TX_n, \ldots, TX_N\}$, respectively. Each M2M-TX is equipped with energy harvesting and data transmission functionalities. A harvest-then-transmit protocol [43] is employed, under which each time block is further divided into two slots: the energy harvesting slot and the data transmission slot. In the energy harvesting slot, M2M-TXs harvest the energy from ambient radio frequency (RF) signals. In the data transmission slot, the stored and harvested energy is utilized by M2M-TXs to transmit data signals to their corresponding M2M-RXs.

We consider the downlink energy harvesting and spectrum reusing scenario due to the following three reasons. First, the transmission power released from the BS in the downlink transmission is more enormous compared to that of CUs, which benefits M2M-TXs with harvesting more energy. Second, the transmission power of M2M-TXs is smaller compared to that of the BS, which can alleviate the QoS degradation of CUs due to the negligible interference brought from M2M-TXs to CUs. Third, although the interference from the BS is strong, the QoS requirement of M2M communication can be satisfied due to the proximate transmission between M2M-TXs and M2M-RXs. The QoS requirement of M2M communication has also been considered during the optimization process. Similar downlink spectrum reusing models have also been employed in several previous works such as [44, 45]. The time allocation decision is defined as follows.

© Springer Nature Switzerland AG 2021
Z. Zhou et al., *Green Internet of Things (IoT): Energy Efficiency Perspective*,
Wireless Networks, https://doi.org/10.1007/978-3-030-64054-5_3

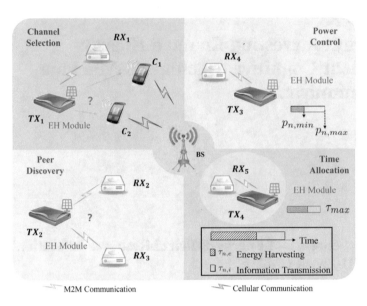

Fig. 3.1 The EH-CM2M communication scenario underlying a single cellular network

Definition 3.1 (Time Allocation Decision) Time allocation here refers to process of deciding the proportion of time spent in energy harvesting and data transmission, respectively. In particular, the total duration of a time slot is denoted by τ_{max}, and the slot durations of energy harvesting and data transmission are defined as $\tau_{n,e}$ and $\tau_{n,i}$, respectively. τ_{max} is a fixed value (or a constant) [46, 47], while $\tau_{n,e}$ and $\tau_{n,i}$ are variables that have to be optimized. Hence, we have $\tau_{n,e} + \tau_{n,i} \leq \tau_{max}$.

Remark 3.1 $\tau_{n,e} = 0$ represents the scenario without energy harvesting, where TX_n transmits data by only using the stored energy. This scenario happens when the following three conditions are satisfied: (1) the energy efficiency gain of energy harvesting is less than that of data transmission; (2) the stored energy is sufficient to support data transmission, i.e., the energy causality constraint is satisfied; (3) the stored energy is sufficient to meet the QoS requirement. Otherwise, if either of the three conditions is unsatisfied, $\tau_{n,e} = 0$ is infeasible.

On the other hand, $\tau_{n,i} = 0$ represents that TX_n only harvests energy and stops data transmission. This scenario happens when either of the following three conditions is valid: (1) considering the intermittent and delay tolerant traffic characteristics of M2M communication, TX_n has no data for transmission in the current slot; (2) TX_n does not have enough energy for data transmission; (3) TX_n has enough energy for data transmission, but does not have enough energy to meet the QoS requirement. Otherwise, if all of the above three conditions are invalid, TX_n can simply increase $\tau_{n,i}$ to achieve a nonzero energy efficiency since $\tau_{n,i} = 0$ is not the optimal solution.

Finally, if TX_n has data for transmission but either the energy causality constraint or the QoS requirement is required, then the arrived data can be stored in a buffer for later transmission. We do not consider packet loss caused by finite buffer size due to the following two reasons. First, the buffer is usually large enough to store a large amount of machine-type data and its capacity continues to grow with the development of hardware. Second, the price of buffer is much cheaper than that of the battery. It is easier to install a large-capacity buffer than a large-capacity battery. Nevertheless, the data overflowing or loss problem caused by limited energy or buffer size is out of the scope of this chapter, and will be considered in our future work.

In the data transmission slot, each M2M-TX has to discover the corresponding M2M-RX, which is henceforth referred to as peer discovery, and determine the communication channel to be reused (i.e., channel selection). In the peer-discovery procedure, an M2M-TX can operate in three modes: (i) the M2M-TX directly transmits data to the BS; (ii) the M2M-TX transmits data to local data aggregators (LDAs) which are densely deployed in the proximity of M2M devices to reduce transmission distance and enhance cell coverage; and (iii) the M2M-TX transmits data to a neighboring M2M device in a peer-to-peer (P2P) fashion [48]. In this work, we focus on the second and third modes, since the first mode is not suitable for the scenario of downlink spectrum reusing. Particularly, if the M2M-TX needs to transmit data to a specific M2M-RX, it does not need to discover the corresponding M2M-RX through the peer-discovery procedure. For the N M2M-TXs, we assume that there are at most M M2M-RXs available, which include both the LDAs and M2M receivers. The set of M M2M-RXs is denoted by $\mathcal{RX}_M = \{RX_1, \ldots, RX_m, \ldots, RX_M\}$.

In the channel selection procedure, CUs are regarded as the primary users, which have the higher priority to use the licensed frequency bands. On the other hand, M2M-TXs are regarded as the secondary users. They share the spectrum resources with CUs under the condition that the spectrum reuse will not harmfully degrade the QoS performance of CUs. To be more specific, we assume an orthogonal partitioning of resources, namely, the spectrum resources of cellular networks are partitioned into time-frequency resource blocks (RBs) [49], and each RB is allocated to a single CU. This way, co-channel interference does not exist among CUs. The set of available RBs is defined as $\mathcal{RB}_K = \{RB_1, \ldots, RB_k, \ldots, RB_K\}$. Note that the RB allocated to C_k is denoted by RB_k.

We consider an example that TX_n communicates with RX_m by reusing RB_k allocated to C_k. As a result of downlink spectrum reusing, C_k suffers from the co-channel interference caused by TX_n, and RX_m suffers from the co-channel interference caused by the BS. Here, we focus on the scenario that each RB can be reused by at most one M2M-TX. The discussion of a more complicated scenario where multiple M2M-TXs are allowed to reuse the same RB is out of the scope of this chapter and treated as one of our future works. The peer discovery and channel selection decision are defined as follows.

Definition 3.2 (Peer Discovery and Channel Selection Decision) The joint peer discovery and channel selection decisions are denoted by an $N \times M \times K$ matrix $\mathbf{S}_{N \times M \times K}$. Here, the (n, m, k)-th element of $\mathbf{S}_{N \times M \times K}$, i.e., $s_{n,m,k}$ is a binary variable. In particular, $s_{n,m,k} = 1$ represents that TX_n and RX_m form an M2M pair by reusing the RB of C_k and $s_{n,m,k} = 0$, otherwise.

In the following subsections, we elaborate the data transmission, energy harvesting, and energy consumption models in details.

3.1.1 Data Transmission Model

We assume that perfect channel state information (CSI) is available during the period of energy harvesting and data transmission. With the adopted channel model, both the slow fading effects and fast fading effects due to pathloss, multipath propagation, and shadowing are taken into account [30, 50]. Thus, the channel power gain of the link from the BS to C_k is given by

$$g_{0,k} = \omega \delta_{0,k} \zeta_{0,k} d_{0,k}^{-\alpha}, \tag{3.1}$$

where ω and α denote the pathloss constant and pathloss exponent, respectively. $\delta_{0,k}$ is the fast-fading gain following an exponential distribution, and $\zeta_{0,k}$ is the slow-fading gain following a log-normal distribution. $d_{0,k}$ is the transmission distance from the BS to C_k.

We assume that TX_n and RX_m form an M2M pair by reusing RB_k, i.e., $s_{n,m,k} = 1$. Under this assumption, the signal to interference plus noise ratio (SINR) of C_k is given by

$$\gamma_k = \frac{p_0 g_{0,k}}{p_n g_{n,k} + N_0}, \tag{3.2}$$

where p_0 and p_n denote the transmission power of the BS and TX_n, respectively. $g_{n,k}$ represents the power gain of the interference link from TX_n to C_k. N_0 is the additive white Gaussian noise.

The received SINR at RX_m for TX_n when reusing RB_k is given by

$$\gamma_{n,m,k} = \frac{p_n g_{n,m,k}}{p_0 g_{0,m,k} + N_0}, \tag{3.3}$$

where $g_{n,m,k}$ denotes the channel power gain from TX_n to RX_m, and $g_{0,m,k}$ denotes the channel power gain from the BS to RX_m.

The achievable spectrum efficiency (defined as bits/Hz) of the M2M pair (TX_n, RX_m) during $\tau_{n,i}$ is given by

$$R_n = \tau_{n,i} \log_2 (1 + \gamma_{n,m,k}). \tag{3.4}$$

3.1.2 Energy Harvesting and Energy Consumption Model

In accordance with the harvest-then-transmit model [43, 51], the harvested energy of TX_n during $\tau_{n,e}$ can be calculated as

$$E_n = \tau_{n,e} \lambda_n p_0 g_{0,n,k},$$ (3.5)

where λ_n denotes the energy harvesting efficiency factor of TX_n ($0 < \lambda_n < 1$), and $g_{0,n,k}$ is the channel power gain of the link from the BS to TX_n.

The power consumption of TX_n consists of two parts: the transmission power consumption p_n and the circuit power consumption p_c. Here, p_c is considered as a constant for simplicity, namely, the value of circuit power consumption is the same for all $TX_n \in \mathcal{TX}_N$. Hence, the energy consumed by TX_n for energy harvesting during $\tau_{n,e}$ is calculated as

$$E_n^e = \tau_{n,e} p_c.$$ (3.6)

The energy consumed by TX_n for data transmission during $\tau_{n,i}$ is given by

$$E_n^i = \tau_{n,i} (p_n + p_c).$$ (3.7)

Consequently, the total energy consumption of M2M-TX M_n is given by

$$E_n^{total} = E_n^e + E_n^i = \tau_{n,i} p_n + (\tau_{n,i} + \tau_{n,e}) p_c.$$ (3.8)

The energy efficiency (bits/Joule) of TX_n, which is defined as the ratio of the achievable spectrum efficiency to the total energy consumption [52], is given by

$$\eta_n = \frac{R_n}{E_n^{total}}.$$ (3.9)

Substituting (3.4), (3.6), (3.7), and (3.8) into (3.9), η_n can be rewritten as

$$\eta_n = \frac{\tau_{n,i} \log_2 (1 + \gamma_{n,m,k})}{\tau_{n,e} p_c + \tau_{n,i} (p_n + p_c)}.$$ (3.10)

3.1.3 Energy Efficient Resource Allocation Problem Formulation

In the EH-CM2M networks, the key research challenge is how to jointly optimize channel selection, peer discovery, power control and time allocation from an energy efficiency perspective. The objective is to maximize the total energy efficiency of the

M2M-TXs under numerous practical constraints such as the constraints of energy causality, QoS provisioning, transmission power, spectrum reusing, etc.

The sets of indices are denoted by $\mathcal{N} = \{1, \ldots, n, \ldots, N\}$, $\mathcal{K} = \{1, \ldots, k, \ldots, K\}$, and $\mathcal{M} = \{1, \ldots, m, \ldots, M\}$, respectively. Furthermore, the set of optimization variables is $\{\mathbf{S}_{N \times M \times K}, \mathcal{P}_n, \mathcal{T}_e, \mathcal{T}_i\}$, where $\mathcal{P}_n = \{p_n | n \in \mathcal{N}\}$, $\mathcal{T}_e = \{\tau_{n,e} | n \in \mathcal{N}\}$, and $\mathcal{T}_i = \{\tau_{n,i} | n \in \mathcal{N}\}$. Thus, the energy-efficient joint channel selection, peer discovery, power control and time allocation optimization problem is formulated as follows:

$$\mathbf{P1}: \max_{\mathbf{S}_{N \times M \times K}, \mathcal{P}_n, \mathcal{T}_e, \mathcal{T}_i} \sum_{n=1}^{N} \sum_{m=1}^{M} \sum_{k=1}^{K} s_{n,m,k} \eta_n(p_n, \tau_{n,e}, \tau_{n,i})$$

$$\text{s.t. } C_1: \gamma_k \geq \gamma_{k,min}, \forall k \in \mathcal{K},$$

$$C_2: s_{n,m,k} \gamma_{n,m,k} \geq s_{n,m,k} \gamma_{m,min}, \forall n \in \mathcal{N},$$

$$\forall k \in \mathcal{K}, \forall m \in \mathcal{M},$$

$$C_3: \tau_{n,e} + \tau_{n,i} \leq \tau_{max}, \forall n \in \mathcal{N},$$

$$C_4: E_n^{total} \leq E_n + E_{n,0}, \forall n \in \mathcal{N},$$

$$C_5: p_{n,min} \leq p_n \leq p_{n,max}, \forall n \in \mathcal{N},$$

$$C_6: \tau_{n,e}, \tau_{n,i} \geq 0, \forall n \in \mathcal{N},$$

$$C_7: s_{n,m,k} \in \{0, 1\}, \forall n \in \mathcal{N}, m \in \mathcal{M}, k \in \mathcal{K},$$

$$C_8: \sum_{m \in \mathcal{M}, k \in \mathcal{K}} s_{n,m,k} \leq 1, \forall n \in \mathcal{N},$$

$$C_9: \sum_{n \in \mathcal{N}, k \in \mathcal{K}} s_{n,m,k} \leq 1, \forall m \in \mathcal{M},$$

$$C_{10}: \sum_{n \in \mathcal{N}, m \in \mathcal{M}} s_{n,m,k} \leq 1, \forall k \in \mathcal{K}. \tag{3.11}$$

where C_1 and C_2 represent the QoS requirements of cellular links and M2M links, respectively. Particularly, $\gamma_{k,min}$ and $\gamma_{m,min}$ denote the minimum SINR thresholds of C_k and the M2M pair (TX_n, RX_m) reusing RB_k, respectively. C_3 is the time allocation constraint. C_4 is the energy casuality constraint, which ensures that the energy consumed by TX_n does not exceed the harvested energy and stored energy $E_{n,0}$. C_5 defines the boundary constraints of transmission power, where $p_{n,min}$ and $p_{n,max}$ are the lower and upper bounds of the transmission power. C_6 guarantees that the time allocation variables are nonnegative. C_7 specifies that the joint peer discovery and channel selection indicator is binary. C_8 guarantees that any M2M-TX can be paired with at most one M2M-RX and one RB. If an M2M-TX is pared with more than one M2M-RX or RB, then C_8 will be violated. Similarly, C_9 guarantees that any M2M-RX can be paired with at most one M2M-TX and one RB. C_{10} guarantees that any RB can be paired with at most one M2M pair. In summary, $C_7 \sim C_{10}$ guarantee the one-to-one relationship among the RB, M2M-TX and M2M-RX. To be specific, each M2M-TX can be paired to at most one M2M-RX, and each M2M pair can utilize at most one RB.

3.2 Energy Efficient Joint Channel Selection, Peer Discovery, Power Control and Time Allocation for EH-CM2M Communications

In this section, we introduce the energy-efficient joint channel selection, peer discovery, power control and time allocation algorithm for EH-CM2M communications.

3.2.1 Matching Based Problem Transformation

It is extremely challenging to find a polynomial-time solution for problem **P1** due to the following reasons.

NP-Completeness for Searching All Possible Combinations As introduced in Sect. 3.1.3, **P1** is a mixed integer nonlinear programming (MINLP) problem, where both integer variables: $\{\mathbf{S}_{N \times M \times K}\}$, and continuous variables: $\{\mathcal{P}_n, \mathcal{T}_e, \mathcal{T}_i\}$ have to be jointly optimized. One of the most commonly used optimization techniques, the brute-force searching method [53], has to examine all the possible combinations of solutions in order to find the optimum solution, i.e., it will examine a total of $N \times M \times K$ combinations in the scenario. As the number of M2M-TXs, M2M-RXs, and RBs increases, the brute-force searching method becomes infeasible since the computation complexity increases dramatically.

Combinatorial Nature of Channel Selection and Peer Discovery The channel selection and peer discovery optimization variable $s_{n,m,k}$ in **P1** is a binary value, which makes the original problem combinatorial. Traditional convex optimization approaches become infeasible due to the tight coupling between integer and continuous variables.

Objective Function Nonconcavity The objective function of **P1** is the summation of a series of fractions, which is generally nonconvex. The standard convex optimization methods cannot be applied, and other low-complexity cost-effective algorithms are required.

The channel selection and peer discovery optimization variable $s_{n,m,k}$ leads to a three-dimensional matching among M2M-TXs, M2M-RXs, and RBs. Keeping this in mind, we transform **P1** into a three-dimensional matching problem. Furthermore, the matching dimensionality can be reduced by combining one M2M-RX and one RB together to form a combinational resource pair, which is henceforth referred to as RXRB pair. With M M2M-RXs and K RBs, a total number of $M \times K$ possible combinations can be established. The set of all possible combinations is denoted by $\mathcal{RXRB}_{M \times K} = \{RBRX_{m,k}\}_{m=1,k=1}^{m=M,k=K}$. This way, the three-dimensional matching problem is then reduced into a two-dimensional matching between N M2M-TXs and $M \times K$ RXRB pairs.

The transformed two-dimensioned matching problem is denoted by a triple $(\mathcal{TX}_N, \mathcal{RXRB}_{M \times K}, \mathcal{F}_N)$, where $\mathcal{TX}_N$ and $\mathcal{RXRB}_{M \times K}$ represent the sets of matching participants, and $\mathcal{F}_N$ represents the set of matching preferences. Based on the preference, a stable matching between an M2M-TX in the set $\mathcal{TX}_N$ and an RXRB pair in the set $\mathcal{RXRB}_{M \times K}$ is formed to maximize the energy efficiency. The definition of a two-dimensional one-to-one matching is given below.

Definition 3.3 (One-to-One Matching) Given the set of M2M-TXs $\mathcal{TX}_N$ and the set of RXRB pairs $\mathcal{RXRB}_{M \times K}$, The matching ϕ denotes a one-to-one correspondence from set $\mathcal{TX}_N \cup \mathcal{RXRB}_{M \times K}$ onto set $\mathcal{TX}_N \cup \mathcal{RXRB}_{M \times K} \cup \emptyset$ such that [54]:

1) $\forall T X_n \in \mathcal{TX}_N, \phi(T X_n) \in \mathcal{RXRB}_{M \times K}$ or $\phi(T X_n) = \emptyset$;

2) $|\phi(T X_n)| \leq 1$.

which guarantee that each M2M-TX could only be allocated at most one RXRB pair.

Obtaining the optimal solution to **P1** is therefore comprised of two stages. In the first stage, each M2M-TX is temporally matched to an available RXRB pair, and then jointly optimize the power control and time allocation to derive the corresponding matching preferences. In the second stage, the joint channel selection and peer discovery optimization problem is solved by matching M2M-TXs with RXRB pairs based on the established preferences in the first stage. The detailed solutions are elaborated as follows.

3.2.2 First-Stage Joint Power Control and Time Allocation Optimization

1. Problem Transformation and Decomposition

Assuming that $T X_n$ is temporally paired with $R X R B_{m,k}$, i.e., $s_{n,m,k} = 1$ and $\phi(T X_n) = R X R B_{m,k}$, the maximum achievable energy efficiency of $T X_n$ under this matching is obtained by solving the following joint power control and time allocation problem:

$$\mathbf{P2}: \max_{p_n, \tau_{n,e}, \tau_{n,i}} \eta_n(p_n, \tau_{n,e}, \tau_{n,i}) \mid_{s_{n,m,k}=1}$$

$$\text{s.t. } C_1 \sim C_6. \tag{3.12}$$

It is obvious that the objective function of **P2** is in fractional from, which is nonconvex with respect to the optimization variables, i.e., p_n, $\tau_{n,e}$, $\tau_{n,i}$. The standard convex optimization methods cannot be applied, and the low-complexity nonlinear fractional programming method is employed to solve **P2** [45].

The implementation procedures are elaborated as follows. First of all, we transform **P2** with the fraction-form objective function into a new problem with a parametric subtractive-form objective function according to the following theorem.

Theorem 3.1 *Defining $\hat{q}_{n,m,k}$ as the optimal objective value of (3.12), the optimal solutions $\hat{p}_n$, $\hat{\tau}_{n,i}$, and $\hat{\tau}_{n,e}$ to **P2** can be obtained if and only if*

$$\max_{p_n, \tau_{n,i}, \tau_{n,e}} \tau_{n,i} \log_2 \left(1 + \gamma_{n,m,k}(p_n)\right)$$

$$- \hat{q}_{n,m,k} E_n^{total}(p_n, \tau_{n,e}, \tau_{n,i})$$

$$= \max_{p_n, \tau_{n,i}, \tau_{n,e}} \tau_{n,i} \log_2 \left(1 + \gamma_{n,m,k}(p_n)\right)$$

$$- \hat{q}_{n,m,k} \left(\tau_{n,e} p_c + \tau_{n,i}(p_n + p_c)\right)$$

$$= \log_2 \left(1 + \gamma_{n,m,k}(\hat{p}_n)\right) - \hat{q}_{n,m,k} E_n^{total}(\hat{p}_n, \hat{\tau}_{n,e}, \hat{\tau}_{n,i}) = 0. \tag{3.13}$$

Proof Theorem 3.1 can be approved by exploiting the property of nonlinear fractional programming. A similar proof can be found in [45]. □

Hence, solving (3.13) is equal to solve (3.12). However, it is noticed that the value of $\hat{q}_{n,m,k}$ is still unknown. To address this challenge, we adopt Dinkelbach method to alternatively derive $q_{n,m,k}$, p_n, $\tau_{n,i}$ and $\tau_{n,e}$ in an iterative fashion. Denote the Dinkelbach iteration as t, the optimal power control and time allocation variables at the t-th iteration, i.e., $\hat{p}_n^t$, $\hat{\tau}_{n,e}^t$ and $\hat{\tau}_{n,i}^t$ are derived based on the value of $q_{n,m,k}$ obtained in the $(t-1)$-th iteration, i.e., $q_{n,m,k}^{t-1}$, by solving the following problem:

$$\mathbf{P3}: \max_{p_n^t, \tau_{n,i}^t, \tau_{n,e}^t} \tau_{n,i}^t \log_2 \left(1 + \gamma_{n,m,k}(p_n^t)\right)$$

$$- q_{n,m,k}^{t-1} \left(\tau_{n,e} p_c + \tau_{n,i}(p_n^t + p_c)\right)$$

$$\text{s.t. } C_1 \sim C_6. \tag{3.14}$$

Then, the value of $q_{n,m,k}$ will be updated as

$$q_{n,m,k}^t = \tau_{n,i}^t \log_2(1 + \gamma_{n,m,k}(p_n^t))/E_n^{total}(p_n^t, \tau_{n,e}^t, \tau_{n,i}^t). \tag{3.15}$$

Finally, the Dinkelbach iteration terminates when

$$\hat{\tau}_{n,i}^t \log_2 \left(1 + \gamma_{n,m,k}(\hat{p}_n^t)\right) - q_{n,m,k}^{t-1} E_n^{total}(\hat{p}_n^t, \hat{\tau}_{n,e}^t, \hat{\tau}_{n,i}^t) \le \delta \qquad (3.16)$$

where δ is the maximum tolerance. The solutions obtained in the last iteration are indeed the optimal solutions.

However, as it can be observed that the power control and the time allocation variables are still coupled with each other in terms of $p_n \tau_{n,i}$ in **P3**. In order to decouple the two variables, we adopt the alternating optimization (AO) approach [55]. In each AO iteration, the power control suboptimal problem is solved by fixing the variables of the time allocation. Then, the derived optimal power control variables are used to optimize the time allocation variables in the next iteration.

Specifically, denote the AO iteration index as l. By using the optimal time allocation variables obtained from the $(l-1)$-th iteration, i.e., $\hat{\tau}_{n,e}^{t,l-1}$ and $\hat{\tau}_{n,i}^{t,l-1}$, the optimal power control variable at the l-th iteration, i.e., $\hat{p}_n^{t,l}$, can be derived by solving the following power control problem:

$$\mathbf{P4}: \max_{p_n^{t,l}} \hat{\tau}_{n,i}^{t,l-1} \log_2 \left(1 + \gamma_{n,m,k}(p_n^{t,l})\right)$$

$$- \hat{q}_{n,m,k}^{t-1} E_n^{total}(p_n^{t,l}, \hat{\tau}_{n,e}^{t,l-1}, \hat{\tau}_{n,i}^{t,l-1})$$

$$\text{s.t. } C_1, C_2, C_4, C_5. \qquad (3.17)$$

After obtaining $\hat{p}_n^{t,l}$, the optimal time allocation variables at the l-th iteration, i.e., $\hat{\tau}_{n,e}^{t,l}$ and $\hat{\tau}_{n,i}^{t,l}$, can be derived by solving the following time allocation problem:

$$\mathbf{P5}: \max_{\tau_{n,e}^{t,l}, \tau_{n,i}^{t,l}} \tau_{n,i}^{t,l} \log_2 \left(1 + \gamma_{n,m,k}(\hat{p}_n^{t,l})\right)$$

$$- \hat{q}_{n,m,k}^{t-1} E_n^{total}(\hat{p}_n^{t,l}), \tau_{n,e}^{t,l}, \tau_{n,i}^{t,l}$$

$$\text{s.t. } C_3, C_4, C_6. \qquad (3.18)$$

In the following, we elaborate how to solve **P4** and **P5** in details.

2. Power Control Optimization Based on the Lagrange Dual Decomposition

The objection of **P4** is a standard concave function with regards to p_n^t. To employ the Lagrange dual decomposition, we firstly combine the constraints C_1 and C_5 into a simpler constraint as follows:

$$\tilde{C}_1: p_{n,min} \le p_n \le \min \left\{ \frac{p_0 g_{0,k} - \gamma_{k,min} N_0}{\gamma_{k,min} g_{n,k}}, p_{n,max} \right\}, \qquad (3.19)$$

The Lagrangian associated with **P4** at the l-th iteration is given by

$$
\begin{aligned}
&\mathcal{L}(p_n^{t,l}, \xi_m, \rho_n, \beta_n, \theta_n) \\
=& \hat{\tau}_{n,i}^{t,l-1} \log_2\left(1 + \gamma_{n,m,k}(p_n^{t,l})\right) - q_{n,m,k}^{t-1} E_n^{total}(p_n^{t,l}) \\
&+ \xi_m\left(\gamma_{m,min} - \gamma_{n,m,k}(p_n^{t,l})\right) + \rho_n\left(E_n^{total}(p_n^{t,l}) - E_n - E_{n,0}\right) \\
&+ \theta_n\left(p_n^{t,l} - \min\{\frac{p_0 g_{0,k} - \gamma_{k,min} N_0}{\gamma_{k,min} g_{n,k}}, p_{n,max}\}\right) \\
&+ \beta_n(p_{n,min} - p_n^{t,l}),
\end{aligned}
\tag{3.20}
$$

where ξ_m and ρ_n are the vectors of Lagrange multipliers corresponding to constraints C_2 and C_4, respectively, and the vectors of Lagrange multiplier associated with constraint $\tilde{C}_1$ are defined as β_n and θ_n.

By exploiting the Lagrange dual decomposition, (3.20) is decomposed as follows:

$$
\min_{(\xi_m, \rho_n, \beta_n, \theta_n \geq 0)} \max_{(p_n^{t,l})} \mathcal{L}(p_n^{t,l}, \xi_m, \rho_n, \beta_n, \theta_n)
\tag{3.21}
$$

The inner power control problem can be solved by setting the first-order derivative of $\mathcal{L}(p_n^{t,l}, \xi_m, \rho_n, \beta_n, \theta_n)$ with regards to $p_n^{t,l}$ as zero. The optimal value $\hat{p}_n^{t,l}$ can be calculated as

$$
\hat{p}_n^{t,l} = \left[\frac{1}{\beta_n - \theta_n - (q_{n,m,k}^{t-1} - \rho_n)\hat{\tau}_{n,i}^{t,l-1}} \right. \\
\left. \frac{\hat{\tau}_{n,i}^{t,l-1} \log_2 e(p_0 g_{0,m} + N_0)^2}{g_{n,m,k}\xi_m + (p_0 g_{0,m} + N_0)} - \frac{N_0}{g_{n,m,k}} - \frac{p_0 g_{0,m}}{g_{n,m,k}} \right]^+
\tag{3.22}
$$

where $[a]^+ = \max\{0, a\}$. Then, the Lagrange multipliers in the outer minimization problem can be updated by using the subgradient method [50].

Remark 3.2 Equation (3.22) shows a water-filling algorithm, where the water level is in inverse proportion to ξ_m, ρ_n, β_n, and is positive proportion to θ_n, $q_{n,m,k}^{t-1}$.

3. Time Allocation Optimization Based on Linear Programming

Upon $\hat{p}_n^{t,l}$ is obtained, it is then utilized to derive the optimal time allocation variables at the l-th iteration, i.e., $\hat{\tau}_{n,e}^{t,l}$ and $\hat{\tau}_{n,e}^{t,l}$. It is noted that **P5** is a linear programming problem with regards to $\tau_{n,e}^t$ and $\tau_{n,i}^t$. Thus, the corresponding optimal solution can be acquired by exploiting the well-knwon simplex method [56].

To further reduce the computational complexity, the following theorem is introduced by investigating the problem structure of **P5**.

Algorithm 3.1 Preference list construction

1: **Input** : $\mathcal{TX}_N, \mathcal{RX}_M, \mathcal{RB}_K$.
2: **Output** : $\mathcal{F}_N$.
3: **for** $TX_n \in \mathcal{TX}_N$ **do**
4:   **for** $RX_m \in \mathcal{RX}_M$ **do**
5:     **for** $RB_k \in \mathcal{RB}_K$ **do**
6:       **Initialization** : set $t = 1, l = 1, s_{n,m,k} = 1$, initialize $\hat{q}_{n,m,k}^{t-1}, \hat{\tau}_{n,e}^{t,l-1}, \hat{\tau}_{n,i}^{t,l-1}, \hat{p}_n^{t,l-1}$,
        and maximum tolerance δ and Γ.
7:       Solve the power control subproblem **P4** for given $\hat{\tau}_{n,e}^{t,l-1}, \hat{\tau}_{n,i}^{t,l-1}, \hat{q}_{n,m,k}^{t-1}$ to obtain $\hat{p}_n^{t,l}$.
8:       Solve the time allocation subproblem **P5** for given $\hat{p}_n^{t,l}, q_{n,m,k}^t$ to obtain $\hat{\tau}_{n,e}^{t,l}$ and $\hat{\tau}_{n,i}^{t,l}$.
9:       **if** $|\hat{p}_n^{t,l} - \hat{p}_n^{t,l-1}| > \Gamma, |\hat{\tau}_{n,e}^{t,l} - \hat{\tau}_{n,e}^{t,l-1}| > \Gamma, |\hat{\tau}_{n,i}^{t,l} - \hat{\tau}_{n,i}^{t,l-1}| > \Gamma$ **then**
10:        $l = l + 1$, return to Line 7.
11:      **else**
12:        **Set** $\hat{p}_n^t = \hat{p}_n^{t,l}, \hat{\tau}_{n,e}^t = \hat{\tau}_{n,e}^{t,l}, \hat{\tau}_{n,i}^t = \hat{\tau}_{n,i}^{t,l}$.
13:        **if** $\hat{\tau}_{n,i}^t \log_2\left(1 + \gamma_{n,m,k}(\hat{p}_n^t)\right) - \hat{q}_{n,m,k}^{t-1} E_n^{total}(\hat{p}_n^t, \hat{\tau}_{n,e}^t, \hat{\tau}_{n,i}^t) \le \delta$ **then**
14:          Calculate $\hat{U}_n|_{\phi(TX_n)=RXRB_{m,k}}$ for obtained $\hat{\tau}_{n,e}^t, \hat{\tau}_{n,i}^t$, and $\hat{p}_n^t$ according to
            (3.23).
15:          Sort every $RXRB_{m,k} \in \mathcal{RXRB}_{M,K}$ in a descending order based on the
            obtained $\hat{U}_n|_{\phi(TX_n)=RXRB_{m,k}}$ to form $\mathcal{F}_n$.
16:        **else**
17:          Update $\hat{q}_{n,m,k}^t$ according to (3.15),
18:          $t = t + 1$, return to Line 6.
19:        **end if**
20:      **end if**
21:    **end for**
22:   **end for**
23: **end for**
24: Construct $\mathcal{F}_N$

Theorem 3.2 *At the l-th iteration, the optimal solutions of problem* **P5**, *i.e.,* $\hat{\tau}_{n,e}^{t,l}$ *and* $\hat{\tau}_{n,i}^{t,l}$, *must satisfy the following condition:* $\hat{\tau}_{n,e}^{t,l} + \hat{\tau}_{n,i}^{t,l} = \tau_{max}$.

Proof Contradiction is utilized to verify the optimal condition. $\{\hat{p}_n^{t,l}, \hat{\tau}_{n,e}^{t,l}, \hat{\tau}_{n,i}^{t,l}\}$ is supposed as the optimal solution which satisfies $\hat{\tau}_{n,e}^{t,l} + \hat{\tau}_{n,i}^{t,l} < \tau_{max}$, that is to say, time is not fully utilized and there is still time left. Similarly, another solution $\{\tilde{p}_n^{t,l}, \tilde{\tau}_{n,e}^{t,l}, \tilde{\tau}_{n,i}^{t,l}\}$ can be construct, where $\tilde{p}_n^{t,l} = \hat{p}_n^{t,l}, \tilde{\tau}_{n,e}^{t,l} + \tilde{\tau}_{n,i}^{t,l} = \tau_{max}$, and $\tilde{\tau}_{n,e}^{t,l}/\tilde{\tau}_{n,i}^{t,l} = \hat{\tau}_{n,e}^{t,l}/\hat{\tau}_{n,i}^{t,l}$, which ensures that the constraints still stand and another feasible solution of **P5** can be obtained. However, this contradicts the original hypothesis. Thus, we must have $\hat{\tau}_{n,e}^{t,l} + \hat{\tau}_{n,i}^{t,l} = \tau_{max}$. $\square$

Based on **Theorem** 3.2, the optimization variables of **P5** can be reduced to only $\tau_{n,e}^{t,l}$ by substituting $\tau_{n,i}^{t,l}$ with $\tau_{max} - \tau_{n,e}^{t,l}$. Hence, rendering that the original time allocation problem turns into a one-dimensional linear optimization problem.

3.2.3   Preference List Construction

In the two-dimensional matching, each M2M-TX has to obtain the respective preferences towards the RXRB pairs. Denote $\mathcal{G}_M = \{G_1, \ldots, G_m, \ldots, G_M\}$ and $\mathcal{G}_K = \{G_1, \ldots, G_k, \ldots, G_K\}$ as the price sets of M2M-RXs and RBs, respectively. For instance, G_m can be considered as the matching cost of RX_m. If any M2M-TX wishes to be matched with RX_m, then it must bear the cost G_m. Thus, the preference of TX_n towards RXRB $RXRB_{m,k}$ is defined as the difference between the achievable maximum energy efficiency under the matching $\phi(TX_n) = RXRB_{m,k}$, i.e., $s_{n,m,k} = 1$, and the total matching cost, which is calculated as

$$U_n \mid_{\phi(TX_n)=RXRB_{m,k}} = \eta_n^* \mid_{\phi(TX_n)=RXRB_{m,k}} -G_m - G_k, \qquad (3.23)$$

where the maximum energy efficiency $\eta_n^* \mid_{\phi(TX_n)=RXRB_{m,k}}$ can be obtained by solving **P2**. To simplify the establishment process of preference list, the initial value for any $G_m \in \mathcal{G}_M$ or $G_k \in \mathcal{G}_K$ can be set as zero. Then, the value of which will be updated during the pricing-based iterative matching process.

We use the binary preference relation "$\succ$" to compare the preferences [57]. For example, $RXRB_{m,k} \succ_{TX_n} RXRB_{m',k'}$, represents that TX_n prefers $RXRB_{m,k}$ to $RXRB_{m',k'}$, which is given by

$$RXRB_{m,k} \succ_{TX_n} RXRB_{m',k'}$$

$$\Leftrightarrow \hat{U}_n \mid_{\phi(TX_n)=RXRB_{m,k}} > \hat{U}_n \mid_{\phi(TX_n)=RXRB_{m',k'}},$$

$$\forall n \in \mathcal{N}, \forall m, m' \in \mathcal{M}, m \neq m', \forall k, k' \in \mathcal{K}, k \neq k'. \qquad (3.24)$$

$\mathcal{F}_n$ is denoted as the preference list of TX_n towards all the RXRB pairs, which is established by ranking all the $M \times K$ RXRB pairs in accordance with the obtained preferences in a descending order. In the process of two-dimensional matching, N M2M-TXs and $M \times K$ RXRB pairs are matched with each other on the basis of the derived preference lists.

3.2.4   Second-Stage Joint Channel Selection and Peer Discovery Based on Matching

After deriving the preference of all M2M-TXs, the second-stage joint channel selection and peer discovery problem can be solved by utilizing the pricing-based matching algorithm. The core contents of the matching algorithm are the *propose* and the *price increasing* rules, which are illustrated as follows:

Definition 3.4 (Propose Rule) For any M2M-TX $TX_n \in \mathcal{TX}_N$, it proposes to the most preferred RXRB pair which ranks as the top place in its preference list $\mathcal{F}_n$.

Definition 3.5 (Increase Price Rule) Denote the set of price increment for M2M-RXs as $\Delta\mathcal{G}_M = \{\Delta G_1, \ldots, \Delta G_m, \ldots, \Delta G_M\}$, and denote the set of price increment for RBs as $\Delta\mathcal{G}_K = \{\Delta G_1, \ldots, \Delta G_k, \ldots, \Delta G_K\}$.

If $RXRB_{m,k}$ has received more than one proposals from M2M-TXs, then the prices of RX_m and RB_k are increased as

$$G_m = G_m + \Delta G_m,$$
$$G_k = G_k + \Delta G_k. \tag{3.25}$$

A pricing-based iterative matching algorithm will be carried out to RXRB pairs with M2M-TXs. The detailed implementation procedure is elaborated as follows.

Phase 1: *Matching Preference Initialization*
- Calculate the preference list $\mathcal{F}_n$ for each M2M-TXs $TX_n \in \mathcal{TX}_N$.
- Initialize Ψ as an empty set. Ψ is denoted as the set of RXRB pairs which are proposed by more than one M2M-TX. Obviously, $\Psi = \emptyset$ at the beginning.
- Set $G_m = 0$ and $G_k = 0$ for any M2M-RX $RX_m \in \mathcal{RX}_M$ and RB $RB_k \in \mathcal{RB}_K$, respectively.

Phase 2: *Iterative Matching*
- If $\exists\phi(TX_n) = \emptyset$, perform the propose rule for M2M-TXs.

  – All the M2M-TX which are unmatched propose to their most preferred RXRB pairs in their preference lists.

- If only one M2M-TX proposes to any RXRB pair $RXRB_{m,k} \in \mathcal{RXRB}_{M \times K}$, then they will be matched directly. Otherwise, $RXRB_{m,k}$ will be added into Ψ.
- Perform the increase price rule for RXRB pairs in the set Ψ.

  – Each M2M-RX $RX_m \in \mathcal{RX}_M$ and RB $RB_k \in \mathcal{RB}_K$ raises its price by ΔG_m and ΔG_k, respectively.
  – Each M2M-TX which has proposed to $RXRB_{m,k}$ renews its preference list and updates its proposal accordingly. As the price of $RXRB_{m,k}$ is increased by $\Delta G_m + \Delta G_k$, some M2M-TXs which used to propose to $RXRB_{m,k}$ will give up because $RXRB_{m,k}$ is no longer their most preferred RXRB pair.

- The increase price rule is continuously carried out until only one M2M-TX proposes to pair $RXRB_{m,k}$.
- Update set Ψ

  – Remove the RXRB pairs which are proposed by only one M2M-TX from Ψ.

  Repeat the above processes iteratively.
  Until each M2M-TX $TX_n \in \mathcal{TX}_N$ has been matched with an RXRB pair or no available RXRB pair could be provided to the unmatched M2M-TXs.

Algorithm 3.2 The pricing-based matching algorithm

1: **Input** : $\mathcal{TX}_N, \mathcal{RX}_M, \mathcal{RB}_K, \Delta\mathcal{G}_M, \Delta\mathcal{G}_K$;
2: **Output** : ϕ;
3: **Initialization** : Every M2M-TX $TX_n \in \mathcal{TX}_N$ builds its preference list based on **Algorithm** 3.1;
4: Set $\phi(TX_n) = \emptyset, \forall TX_n, \Psi = \emptyset$;
5: **while** $\exists \phi(TX_n) = \emptyset$ **do**
6:     **for** $TX_n \in \mathcal{TX}_N$ **do**
7:         Implement the *propose* rule: each M2M-TX $TX_n \in \mathcal{TX}_N$ proposes to the top ranking RXRB pair $RXRB_{m,k}$ in its preference list $\mathcal{F}_n$;
8:     **end for**
9:     **if** $\Psi = \emptyset$ **then**
10:         Match M2M-TXs with the requested RXRB pairs directly;
11:     **else**
12:         **for** every $RXRB_{m,k} \in \Psi$ **do**
13:             Implement the *price increasing* rule: the price of RX_m and RB_k will be increased with an amount ΔG_m and ΔG_k, respectively, then the TX_n will update the corresponding preference lists $\mathcal{F}_n$. The *price increasing* rule is continuously carried out until only one M2M-TX proposes to pair $RXRB_{m,k}$;
14:         **end for**
15:     **end if**
16:     Update $\mathcal{F}_n$ by deleting the matched RXRB pairs.
17: **end while**

Phase 3: *Channel Selection Implementation*

Each M2M-TX selects the corresponding channel of CU by accessing RX_m in accordance with the matching result acquired in **Phase 2**. Considering $\phi(TX_n) = RXRB_{m,k}$, TX_n will select the channel of C_k by accessing RX_m to utilize the spectrum resource.

Remark 3.3 It is noteworthy that any M2M-TX $TX_n \in \mathcal{TX}_N$, which can not satisfy the constraint of total time duration, or the energy constraint, can not establish the preference list $\mathcal{F}_n$ in spite of the preference.

3.3 Performance Results and Discussions

In this section, we validate the energy-efficient resource allocation algorithm through simulations. The simulation parameters are summarized in Table 3.1 [58]. The cell radius is 200 m. The locations of CUs are randomly distributed throughout the cell, while the M2M-TXs and M2M-RXs are randomly distributed in a circular area with a radius of $d_{max} = 25$ meters, which is the maximum distance allowed to establish an M2M link. We adopt the Monte Carlo approach, and the results are averaged over 1000 rounds of simulations. The algorithm is compared with two heuristic algorithms, i.e., matching with power control and fixed time allocation, which is developed in [59], and matching with time allocation and fixed power. In

Table 3.1 Simulation parameters

Simulation parameters	Value
Cell radius r (m)	200
Pathloss constant ω	0.01
Fast-fading gain δ	1
Slow-fading gain ζ (dB)	8
Pathloss exponent α	4
Energy harvesting efficiency factor λ	0.9
SINR threshold $\gamma_{k,min}$ (dB)	5
SINR threshold $\gamma_{m,min}$ (dB)	5
Maximum transmission distance of M2M pairs d_{max} (m)	5–35
One-sided power spectral density of the additive white Gaussian noise N_0 (dBm)	−114
Total mximum time τ_{max} (s)	1
Energy storage E_0 (J)	0.5
Transmission power range p_{min} and p_{max} (dBm)	0–23
Circuit power consumption p_c (dBm)	20

particular, the last algorithm always allocates the maximum transmission power to M2M-TXs.

3.3.1 Improve Average Energy Efficiency of M2M-TXs

A snapshot of $N = M = 10$, $K = 12$ and $d_{max} = 25$ m is shown in Fig. 3.2. Figure 3.4 shows the average energy efficiency performance of M2M-TXs versus the maximum transmission distance d_{max} of M2M pairs with $N = M = K = 5$. Simulation results demonstrate that the algorithm can achieve superior performance compared with the other two heuristic algorithms. When $d_{max} = 25$ m, the performance of the algorithm outperforms those two heuristic algorithms by 36.46% and 18.17%, respectively. Moreover, it is apparently that the energy efficiency performance decreases monotonically with the allowed transmission distance. The reason is that as the transmission distance increases, M2M-TXs have to increase the transmission power in order to meet the QoS requirement, which results in lower energy efficiency.

Figure 3.3 shows the relationship among energy efficiency, power control and time allocation. The energy harvesting ratio is defined as the percentage of the time utilized to harvest energy. For example, a ratio of 0.25 represents that 25% of the time is spent in energy harvesting. When the amount of stored energy is sufficient, the energy efficiency performance degrades as the energy harvesting ratio increases. The reason can be inferred from (3.10), i.e., the first-order derivative of η_n with regards to $\tau_{n,i}$ is positive. For the same energy harvesting ratio, it is observed that the energy efficiency firstly increases then decreases with the transmission power. Particularly, using the power which is beyond the optimal transmission power leads to rapid degradation of the energy efficiency performance. The reason is that the

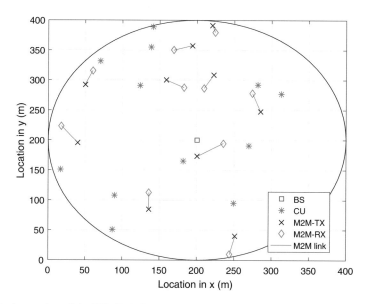

Fig. 3.2 A snapshot of the EH-CM2M network with N M2M-TXs, M M2M-RXs and K CUs ($N = M = 10$, $K = 12$; the cell radius is 200 m)

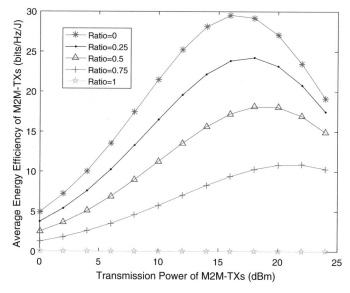

Fig. 3.3 Average energy efficiency of M2M-TXs versus transmission power of M2M-TX ($N = M = K = 5$, $d_{max} = 25$ m)

spectrum efficiency gained by per unit increment of transmission power cannot compensate for the increased energy consumption.

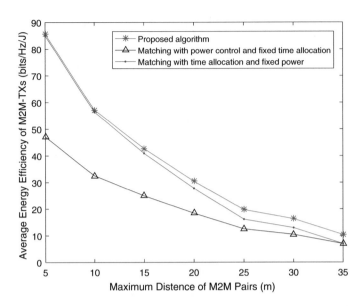

Fig. 3.4 Average energy efficiency of M2M-TXs versus the maximum distance of M2M pairs ($N = M = K = 5$, $d_{max} = 5 \sim 35$ m)

3.3.2   Improve Average Energy Efficiency of M2M Pairs

Figure 3.5 shows the energy efficiency performance versus the number of M2M-RXs. It is noted that the energy performance increases significantly with number of M2M-RXs. The reason is that a large number of M2M-RXs provides a higher probability for the same M2M-TX to be matched with a more preferred RXRB unit, which is also called the *diversity gain*. Furthermore, it is apparently that the algorithm can achieve superior performance compared with the other two heuristic algorithms. When $M = 6$, the energy efficiency performances of the matching with power control and fixed time allocation, and the matching with time allocation and fixed power can achieve up to only 60.90% and 90.34% of the performance achieved by the algorithm. Therefore, the algorithm is able to well exploit the diversity benefits brought by the increasing numbers of M2M-RXs.

Figure 3.6 shows the average energy efficiency of M2M pairs versus the number of M2M-TXs, M2M-RXs, and CUs. The lower limit is defined as the smallest data point that extends to 1.5 times the frame height from the bottom of the box, and the upper limit is defined similarly. The value that exceeds the lower or upper limit of the box is indicated by a red plus sign, which is also known as an outlier. It is apparently that the achieved energy efficiency of the scheme with more number of M2M-RXs is much higher. For instance, when M is increased from 5 to 10, the median value and the maximum value is improved by 4.46% and 5.42%, respectively. The reason

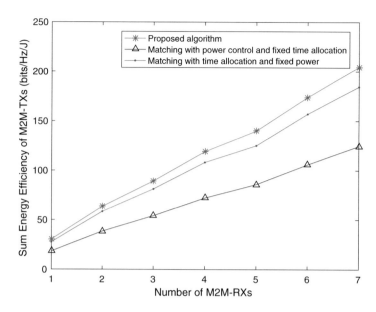

Fig. 3.5 Average energy efficiency of M2M pairs versus number of M2M-RXs ($N = 7$, $M = 1 \sim 7$, $K = 10$, $d_{max} = 25$ m)

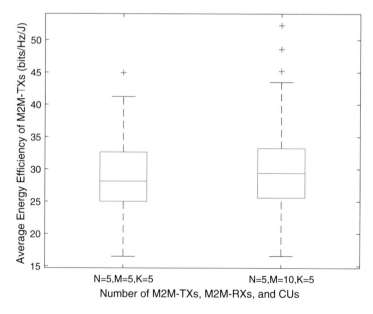

Fig. 3.6 Average energy efficiency of M2M pairs versus number of M2M-TXs, M2M-RXs, and CUs ($d_{max} = 25$ m)

is that the M2M-TX could have a higher probability to be matched with a more preferred RXRB unit from a large number of RBs. Furthermore, it is obviously that the distribution with more M2M-RXs will bring more outliers. The reason lying behind is that the M2M-TX will have more opportunities to choose the adjacent M2M-RX, leading the sharp improvement of the performance of energy efficiency.

In this chapter, we investigated the energy-efficient resource allocation problem for the EH-CM2M communication. We introduced a two-stage energy efficient joint channel selection, peer discovery, power control, and time allocation optimization algorithm by combining AO, nonlinear fractional programming, linear programming and iterative pricing-based matching theory. As shown in Fig. 3.5, simulation results demonstrate that the algorithm can achieve superior performance comparing with other two heuristic algorithms. For instance, the energy efficiency performance of the other heuristic algorithms can achieve up to only 60.90% and 90.34% when $N = 7$, $M = 6$, $K = 10$ comparing with the algorithm, respectively.

Chapter 4
Software Defined Machine-to-Machine Communication for Smart Energy Management in Power Grids

4.1 Framework of Energy-Efficient SD-M2M for Smart Energy Management

This section provides a detailed illustration of the SD-M2M architecture, with a particular emphasis on its technical contributions to complexity reduction, fine granularity resource allocation, and end-to-end QoS guarantee.

4.1.1 Architecture Overview

Figure 4.1 shows the SD-M2M architecture that can be divided into four different planes: the data plane, the control plane, the application plane, and the management and administration plane. The data plane is composed of all of the programmable field equipment and network elements involved in M2M communication, such as sensors, actuators, IEDs, smart meters, gateways, BSs, switches, routers, and so on. These are essential to support autonomous data acquisition and transmission in smart energy management. With the data-control decoupling, the data-plane devices are greatly simplified without the need to understand hundreds of communication protocols.

The control plane consists of a SD-M2M hypervisor and multiple heterogeneous or homogeneous SD-M2M controllers. The virtualization of the physical M2M networks is enabled by inserting an hypervisor between the data-plane devices and the controllers. The hypervisor views and interacts with the data-plane devices through the standard-based southbound interface and slices the abstracted physical infrastructures into multiple isolated virtual M2M networks that are controlled by their respective controllers. The hypervisor also sends the abstraction information to the controllers through the southbound interface. The centralized SD-M2M controller makes decisions on an up-to-date global view of the network state, and

© Springer Nature Switzerland AG 2021
Z. Zhou et al., *Green Internet of Things (IoT): Energy Efficiency Perspective*,
Wireless Networks, https://doi.org/10.1007/978-3-030-64054-5_4

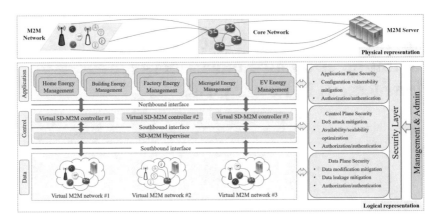

Fig. 4.1 The conceptual architecture of software-defined M2M communication

enables vendor-independent control over the corresponding virtual M2M network from a single logical point. This allows the implementation of fine granularity control policies with enhanced network resource utilization efficiency and QoS provisioning capabilities.

The application plane covers an array of smart energy management applications such as home energy management (HEM), factor energy management (FEM), building energy management (BEM), microgrid energy management (MEM), electric vehicle energy management (EVEM), and so on. With standard-based APIs between the control and application planes, smart energy management applications can explicitly and programmatically communicate their requirements to the respective controllers via the northbound interface, and can thus operate on an abstraction of the M2M networks without being tied to the details of physical infrastructures.

The management and administration plane provides management and access control functions to all the other three planes, i.e., the data plane, the control plane, and the application plane. It covers static tasks such as device setup and management, privacy and security policy configuration, firmware and software updates, performance monitoring, etc. The security layer protects the data plane from various security threats such as flow rule modification, unauthorized access control, side channel attack, and so on. In the control plane, the security layer provides solutions for controller access authorization and authentication, denial of service (DoS) or distributed DoS (DDoS) attack mitigation, controller availability and scalability optimization, etc. Furthermore, security enforcement mechanisms can be implemented to secure the application plane from unauthorized and unauthenticated applications, fraudulent rule insertion, configuration vulnerabilities, and other application-specific security threats.

4.1.2 The Benefits of the SD-M2M

Fine granularity resource allocation in multi-tenant environment. In SD-M2M, physical infrastructures are abstracted from three dimensions of attributes, namely topology, physical device resources, and physical link resources. We focus on how to realize M2M infrastructure abstraction for intelligent service orchestration and resource allocation rather than redesign the concept of network function virtualization. The degree of abstraction for each dimensional attribute can be flexibly controlled by the adjustment of physical resources. Firstly, in the abstraction of network topology, the degree of abstraction relies on the virtual nodes and links. For an instance, a physical topology which represents the layout of connected devices can be either abstracted as an identical virtual topology in the lowest degree of abstraction, or as a single virtual node or link in the highest degree of abstraction. Secondly, the degree of abstraction for physical device resources is dependent on CPU, memory, storage, and other computing resources. Thirdly, in the abstraction of physical link resources, the ability to choose different levels of abstraction is determined by the allocation of link bandwidth, buffers, queues, and so on. Hence, SD-M2M offers granular level of resource allocation in a highly abstracted and automated fashion, and allows the same physical infrastructures to be shared among multiple tenants.

4.2 Software-Defined M2M Communication for Smart Energy Management Applications

We focus on several sub-areas where SD-M2M will play a key role and present how to integrate SD-M2M with different applications in a bottom-up approach. A comprehensive summary of the communication features and critical aspects for smart energy management applications is provide in Table 4.1.

Home Energy Management (HEM)

HEM enables residential energy consumers to be actively involved in the grid operation through intelligent interaction with external environment. Intelligent machines are embedded to collect home appliance operation status, energy consumption, home environment, and home user behaviors for smart HEM. SD-M2M will play a key role in facilitating HEM by shielding vendor-specific details and features of home appliances from application development and system operation. All of the registered M2M devices in a home can be divided into virtual networks with abstracted network, storage, and computing capability, and be managed through standard APIs to deliver home user demand-oriented services in short time.

Building Energy Management (BEM)

The residential and commercial buildings have been estimated to represent approximately half of the total world energy consumption. M2M communications are

Table 4.1 A comprehensive summary of the communication features and critical aspects for smart energy management applications

Application	Communication features	Critical aspects	Benefits of SD-M2M
Home energy management	• Delay tolerant • Periodic/event based • Short range • Low-level priority	• Diverse communication protocols • Massive connection • High random access loads • Small burst traffic	• Reduced cost and complexity • Accelerated innovation • Vendor-independent control
Building energy management	• Delay tolerant • Periodic/event based • Short range • Low-level priority	• Diverse communication protocols • Massive connection • Small burst traffic	• Reduced cost and complexity • Accelerated innovation • Coordinated management • Vendor-independent control
Factory energy management	• Delay sensitive • Periodic/event based • Middle-level priority • Middle range	• Diverse communication protocols • High reliability • Middle-level QoS requirement	• Reduced cost and complexity • Accelerated innovation • Coordinated management • Vendor-independent control
EV energy management	• Delay sensitive • Semi-periodic/event based • Middle-level priority • Middle range	• Mobility management • Random charging/discharging behaviors • High reliability • Middle-level QoS requirement	• Reduced cost and complexity • Coordinated mobility management • Fine granularity resource allocation
Microgrid energy management	• Delay sensitive • Semi-periodic/event based • High-level priority • Middle range	• High reliability • High-level QoS requirement • Multi-tenant environment • Massive connection	• End-to-end QoS guarantee • Fine granularity resource allocation • Coordinated management

critical to collect real-time data of temperature, occupancy behavior, outdoor environment, humidity, illuminance, and electricity price, etc. Smart BEM is realized by dynamically optimizing the energy consumption related to heating, cooling, ventilation, and lighting. SD-M2M provides a comprehensive platform to interact with M2M devices deployed in various building monitoring, control and automation systems, which are usually developed based on different communication protocols. M2M networks in different buildings and systems can be abstracted and integrated into the same virtual network, which provides the benefit of allowing multiple buildings to be remotely managed by a centralized controller.

Factory Energy Management (FEM)

Smart FEM will be a key enabler for the incoming fourth industrial revolution. M2M devices are installed in factory to collect not only energy generation, storage, and consumption data, but also to monitor real-time status of manufacture lines. SD-M2M enables FEM operators to build highly reliable and programmable communication networks for integrating distributed renewable energy sources and energy-saving equipment such as motors and inverters. With the decoupling of data and control planes, the coexistence of energy and manufacturing traffic in the same communication network is supported through physical infrastructure abstraction and centralized resource coordination. Energy consumption improvement points and deterioration factors can be easily identified by interrelating product information with energy information through open and programmable APIs.

Electric Vehicle Energy Management (EVEM)

The massive amounts of data in every aspect of electric vehicles including locations, travel patterns, driver behaviors, battery states, and historical profiles are routinely collected for realizing smart EVEM, which reduces the energy demand-supply imbalance by absorbing excess energy during off-peak hours and discharging the batteries into the grid when needed [60]. SD-M2M with centralized intelligence provides an flexible communication network for coordinated charging and discharging different types of electric vehicles at distributed locations such as residential community, working places, parking lots, and charging stations. For mobility management, seamless handover of electric vehicles from one base station to another can be realized by coordinating network control and orchestrating resource allocation among multiple controllers.

Microgrid Energy Management (MEM)

Microgrid is a small-scale electric power system with co-located distributed energy sources and loads. It can either synchronize to the main grid and operate in grid-connected mode, or operate in island mode by disconnecting both loads and energy sources from the main grid [61]. Hence, MEM provides the benefits of relieving the stress of load-supply imbalance through local consumption of distributed renewable energy sources. In SD-M2M, the physical M2M infrastructure can be abstracted and sliced into distinct virtual networks to support a variety of microgrid energy management functions with diverse communication requirements. Sufficient communication and computing resources should be allocated for the monitoring and control of critical interconnection points, i.e., the points of load connection, common coupling, and distributed energy source connection, in order to support seamless dispatch, scheduling, and control of distributed energy sources. Various stakeholders such as microgrid operator, distributed energy source operator and aggregator, and load aggregator, are allowed to exchange key operating parameters in real-time with the supported coordination among heterogeneous controllers.

4.3 Case Study and Analysis

The case study is divided into two parts. In the first part, we evaluate the capability of SD-M2M for supporting real-time delivery of strictly delay-sensitive data. In the second part, we demonstrate the relationship between SD-M2M penetration rate and performance gain of smart energy management.

When an M2M device attempts to be connected with a BS, it has to randomly select a preamble and send it to the BS via a time-frequency resource block. The BS decodes the received preamble, and sends back a response message. A random access collision occurs if two or more M2M devices happen to select the same resource block, and then each M2M device has to wait for a random period and repeat random access again. Thus, the operation of smart energy management is in danger since critical data cannot be delivered immediately without delay. In particularly, the probability of collision increases dramatically when a massive number of M2M devices attempt to access the network simultaneously.

SD-M2M provides a promising solution to the above challenge through advanced level of resource abstraction and fine granularity resource allocation. The hypervisor slices the physical infrastructure into K distinct virtual M2M networks based on QoS requirements. Without loss of generality, we focus on the k-th$(k = 1, \cdots, K)$ virtual network with N_k M2M devices. Assuming that M_k resource blocks are allocated by the controller, the total number of resource blocks is calculated as $\sum_{k=1}^{K} M_k$. Given $K = 20$ and $M_k = 10$ for $k = 1, 2, \cdots, K$, the total number of required resource blocks is 200. Each M2M device only needs to be aware of the resource blocks allocated to the corresponding virtual network instead of sensing the whole physical network. If the achieved spectrum efficiency cannot meet the specified QoS requirement, more resources can be allocated to this virtual network for improving performance by coordinating resource allocation with other virtual networks. The study of inter-virtual network coordination is left for future study.

The strategy of each M2M device is to decide when to access the network and which resource block to choose. Since random access will be successful if and only if the resource block is idle and is not requested by others, the achievable spectrum efficiency is jointly determined by the number of available resource blocks, the actions of other M2M devices, and the channel quality of the requested resource block. As a result, each M2M device needs to decide whether or not to access the network at each time slot based on the state of resource blocks. Markov decision process (MDP) provides an effective mathematical framework to formulate this category of decision making problems with stochastic process. A standard MDP formulation involves the following elements: state, action, cost function, and state transmission. The system state S is defined as the set of all resource block states. The state transition probability can be modeled as a Poisson process. The action is defined as the probability to access the network, which is relative to the system state. The optimization objective is to maximize the average transmission rate per device over the infinite time horizon. The MDP problem can be broken down into a collection of simpler subproblems and solved one by one via dynamic programming [62]. The algorithm is guaranteed to obtain the optimal performance upon termination. The relative proof can be found in [62] and references therein.

4.3.1   Improve Spectral Efficiency

We compare the algorithm with a base-line greedy algorithm in which each device always requests the resource block with the best channel quality. The results are shown in Fig. 4.2a and b. We consider a virtual network with $N_k = 100$ M2M devices. Figure 4.2a shows the average transmission rate per device with different numbers of resource blocks M_k. The algorithm outperforms the greedy algorithm by more than 300% when $M_k = 10$. The reason is that the reuse gain of resource blocks is fully exploited. In Fig. 4.2b, we fix the total number of resource blocks $M_k = 6$, and change the maximum probability of accessing the network from 12 to 20%. It is shown that with the increase of maximum access probability, the performance degrades dramatically. The reason is that the collision probability increases exponentially as more devices attempt to access the network simultaneously. Nevertheless, the algorithm still outperforms the greedy algorithm under all scenarios.

To evaluate the smart energy management performance, a robust energy scheduling approach in our previous work [60] is employed. Robust energy scheduling allows a distribution-free model of uncertainties, and can efficiently alleviate the negative effect of data uncertainty. The goal is to minimize the generation cost of the gas generator under the constraints of active power balance, active power generation limits, charging and discharging power boundaries, charging demand balance, and spinning reserve. The optimization variables are when to charge and

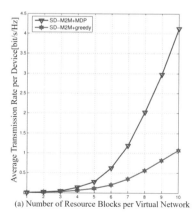
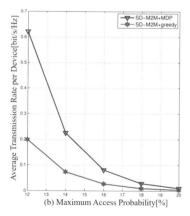

(a) Number of Resource Blocks per Virtual Network   (b) Maximum Access Probability[%]

Fig. 4.2 Spectral efficiency performance: (**a**) average transmission rate per device versus the number of total RBs; (**b**) average transmission rate per device versus the probability of accessing the network

discharge electric vehicles, and the energy output of gas generator. More details of
the robust energy scheduling solution can be found in [60] and references therein.
An electric vehicle cannot be scheduled if the critical data are not delivered on time,
which occurs when either SD-M2M devices are not deployed nor QoS requirement
is violated due to collision. We define the SD-M2M penetration rate as the ratio of
electric vehicles which can be scheduled to the total number of electric vehicles.

4.3.2   Reduce the Total Energy Generation Cost

Figure 4.3a shows the energy supply and demand profiles of electric vehicles, wind
turbines, and local residents for a duration of 24 min. It is noted that the peak load
starts at the sixth minute when the wind power output is low and the charging
demand of electric vehicles is high. Figure 4.3b shows the total energy generation
cost versus the SD-M2M penetration rate. It is obvious that there is a positive
correlation between cost reduction and penetration rate. For an instance, the cost
is reduced by 65% when the SD-M2M penetration rate is increased from 20 to
100%. It is interesting to note that the increment of SD-M2M penetration rate
converts the exponential growth pattern of cost into linear grown pattern. Based
on the delay-sensitive mission-critical data delivered by SD-M2M, the peak load
can be efficiently shifted by charging electric vehicles to absorb renewable energy
during off-peak hours and discharging to produce energy during peak hours.

   In this chapter, we introduced a new software-defined M2M framework for
emerging smart energy management applications. We introduced the current
research progress on integrating SDN with M2M. Then, the design principle
of the SD-M2M architecture was presented, and the technical contributions to

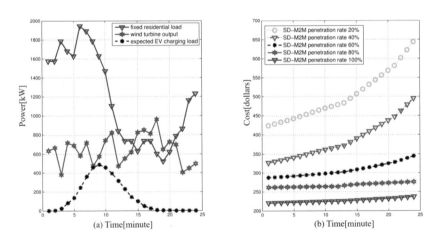

Fig. 4.3 The smart energy management performance: (**a**) the energy supply and demand profiles;
(**b**) the relationship between the total energy generation cost and the SD-M2M penetration rate

cost and complexity reduction, end-to-end QoS guarantee, and fine granularity resource allocation were elaborated in details. We also classified smart energy management applications into several classes based on operation domains, and provided a detailed treatment on how to integrate SD-M2M with each class of application. A case study was conducted in electric vehicle network to demonstrate the performance gains brought by SD-M2M in both spectral efficiency and energy management.

Chapter 5
Energy-Efficient M2M Communications in for Industrial Automation

5.1 Framework of Energy-Efficient M2M Communications

The system model is presented in Fig. 5.1. We consider a single cell with a centralized BS. In the cell, there are J delay-tolerant MTC devices, and K delay-sensitive MTC devices connecting to the BS simultaneously, the sets of which are denoted as $\mathcal{DT}_J = \{DT_1, \cdots, DT_j, \cdots, DT_J\}$ and $\mathcal{DS}_K = \{DS_1, \cdots, DS_k, \cdots, DS_K\}$, respectively. The sets of indices are denoted as $\mathcal{J} = \{1, \cdots, j, \cdots, J\}$, and $\mathcal{K} = \{1, \cdots, k, \cdots, K\}$, respectively. The uplink data transmission from MTC devices to the BS involves two stages: access control and resource allocation.

In the stage of access control, the BS can impose some limitations on the number of devices that are allowed to access to the network. A simple mechanism is the access class barring (ACB) scheme [63]. Initially, the BS broadcasts an ACB barring factor B_0 within the range [0, 1] to all the delay-tolerant MTC devices. After receiving B_0, any MTC device $DT_j \in \mathcal{DT}_J$ will generate a random access number uniformly within [0, 1], e.g., $\hat{B}_j \in [0, 1]$. Device DT_j is allowed to connect to the BS if and only if $\hat{B}_j \leq B_0$. Since $\hat{B}_j$ is uniformly distributed within [0, 1], the probability of $\hat{B}_j \leq B_0$ is exactly B_0. In other words, B_0 is also the access probability. For example, $B_0 = 0.8$ represents that the probability that DT_j is allowed to access is 80%.

Based on ACB, we further introduce a contract-based incentive mechanism to motivate some delay-tolerant MTC devices to postpone their access demands, so that the peak-time access demands can be reduced. Specifically, the BS designs a contract for diverse types of MTC devices with different maximum tolerable delay, which contains a great variety of contract items. In each contract items, the relationship between the performances, i.e., the waiting time, and the reward, i.e., the improvement of access probability, is specified. Then, the BS broadcasts all the contract items, and each delay-tolerant MTC device will choose a contract item to

© Springer Nature Switzerland AG 2021
Z. Zhou et al., *Green Internet of Things (IoT): Energy Efficiency Perspective*,
Wireless Networks, https://doi.org/10.1007/978-3-030-64054-5_5

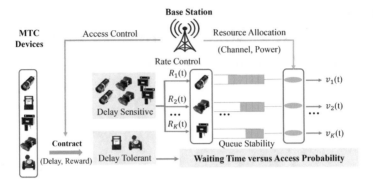

Fig. 5.1 System model

maximize its payoff. The details of how to design and optimize the contract items are described in Sect. 5.2.

In the stage of resource allocation, we consider the K delay-sensitive MTC devices, i.e., $\mathcal{DS}_K$, which are connected to the BS. We adopt a time-slot model. At the t-th slot, the applications require that $R_k(t)$ bits of data should be sensed by device DS_k, which are firstly stored in a buffer of DS_k, i.e., queue k, before being sent to the BS. Let $v_k(t)$ denote the physical-layer transmission rate for queue k at time slot t. In other words, $R_k(t)$ and $v_k(t)$ represent the amount of data entering or leaving the queue. Particularly, $R_k(t)$ and $v_k(t)$ also specify how much data should be sent to the BS from the perspectives of application layer and physical layer, respectively. Let $Q_k(t)$ denote the data backlog, i.e., the amount of data buffered at queue k. We introduce a long-term cross-layer online resource allocation algorithm to optimize $R_k(t)$ and $v_k(t)$ while simultaneously guaranteeing $Q_k(t)$ is stable, which is described in Sect. 5.3.

5.2 Contract-Based Incentive Mechanism Design for Access Control

In this section, we introduce the contract-based incentive mechanism for access control.

5.2.1 MTC Type Modeling

Delay-tolerant MTC devices can be classified into different types based on their delay tolerance. Intuitively, a device with a higher delay tolerance can wait longer and thus contribute more to flatten the peak demand. Therefore, the definition of device type can be determined from the perspective of maximum tolerable delay.

Definition 5.1 (MTC Device Type) The J delay-tolerant MTC devices can be sorted based on the maximum tolerable delay and classified into M types. Denoting the set of device types and the set of delay tolerance capability as $M = \{1, \cdots, m, \cdots, M\}$ and $\Theta = \{\theta_1, \cdots, \theta_m, \cdots, \theta_M\}$, respectively, we have $1 < \cdots < m < \cdots < M$ and $\theta_1 < \cdots < \theta_m < \cdots < \theta_M$.

Let $D_{j,max}$ represent the maximum tolerable delay of device DT_j. $D_{j,max}$ falls into a continuous closed interval $[D_{min}, D_{max}]$, where D_{min} and D_{max} represent the lower and upper bound of $D_{j,max}$, respectively. Next, this interval $[D_{min}, D_{max}]$ is divided into M subintervals of equal length. Denoting θ_m as the lower bound of the m-th subinterval, a device DT_j belongs to type m if $\theta_m \leq D_{j,max} < \theta_{m+1}$.

Next, we introduce the definition of information asymmetry.

Definition 5.2 (Information Asymmetry) Due to the costs of signalling overhead and privacy concerns, the BS does not know each device's type. The specific types of MTC devices are only known by themselves, the information of which is asymmetric.

Nevertheless, the BS can estimate the statistical knowledge of device types based on historical data. Denoting the probability of device DT_j belonging to type m as $P_{j,m}$, we have $\sum_{m=1}^{M} P_{j,m} = 1$, $\forall j \in \mathcal{J}, \forall m \in \mathcal{M}$. By assuming that this probability distribution is *i.i.d.* with regards to devices, the first subscript of $P_{j,m}$ can be removed, and it can be simplified as P_m. It is assumed that the BS knows the knowledge of P_m.

5.2.2   Contract Formulation

For type m MTC device, the BS can require it to postpone its access demand by a period of T_m. Accordingly, the BS will increases its access probability from B_0 to $B_0 + B_m$. Here, B_m is the reward offered by the BS, i.e., the increased value of the access probability. This performance-reward association can be specified via a contract item, which is defined as follows.

Definition 5.3 (Contract) The contract is denoted as the set $C = \{(T_m, B_m), \forall m \in M\}$. The performance-reward association (T_m, B_m) denotes the contract item designed for type m device.

By signing the contract item (T_m, B_m) with type m MTC device, the utility of the BS is given by

$$U_{BS}(T_m, B_m) = \gamma_0 \log_2(T_m) - \beta_0 B_m, \tag{5.1}$$

where γ_0 is the profit coefficient. $\log_2(T_m)$ is the profit gained from the deferred access time T_m, i.e., the waiting time. The logarithmic function is utilized to represent that the marginal profit declines gradually over T_m. β_0 is the unit cost of increasing the ACB barring factor.

Given the J delay-tolerant MTC devices, the expected utility of the BS is calculated as

$$U_{BS}(\{T_m\}, \{B_m\}) = J \sum_{m=1}^{M} P_m \left(\gamma_0 \log_2(T_m) - \beta_0 B_m \right). \tag{5.2}$$

The utility function of type m MTC device which accepts the contract item (T_m, B_m) is given by

$$U_m(T_m, B_m) = \theta_m B_m - \gamma T_m, \tag{5.3}$$

where γ is the unit cost of waiting time and $\theta_m B_m$ represents the profit.

The objective of the BS is to maximize its expected utility with asymmetric information. The contract optimization problem is formulated as

$$\mathbf{P1}: \max_{\{T_m\}, \{B_m\}} U_{BS}(\{T_m\}, \{B_m\}),$$

$$\text{s.t.} \ C_1: \theta_m B_m - \gamma T_m \geq 0, \forall m, (IR)$$

$$C_2: \theta_m B_m - \gamma T_m \geq \theta_m B_{m'} - \gamma T_{m'}, \forall m, m', (IC)$$

$$C_3: 0 \leq B_1 < \cdots < B_m < \cdots < B_M,$$

$$C_4: T_m \leq \theta_m, \forall m, \tag{5.4}$$

where C_1 is the IR constraint, which represents that type m MTC device will get a nonnegative payoff if it selects the contract item (T_m, B_m). C_2 denotes the IC constraint, which guarantees that the type m MTC device will get the maximum payoff if and only if it selects the contract item (T_m, B_m) designed for its own type. C_3 is monotonicity constraint, which specifies that the reward increases monotonically with the device type. C_4 is the upper bound of waiting time T_m.

5.2.3 Contract Optimization

First, we can derive the following properties:

Theorem 5.1 Contract feasibility: The contract $C = \{(T_m, B_m), \forall m \in \mathcal{M}\}$ is feasible if and only if all the following conditions are satisfied:

- $0 \leq B_1 < \cdots < B_m < \cdots < B_M$ and $0 \leq T_1 < \cdots < T_m < \cdots < T_M$;
- $\theta_1 B_1 - \gamma T_1 \geq 0$;

- For any $m \in \{2, \cdots, M\}$, $\gamma T_{m-1} + \theta_{m-1}(B_m - B_{m-1}) \le \gamma T_m \le \gamma T_{m-1} + \theta_m(B_m - B_{m-1})$.

Proof The detailed proof of Theorem 5.1 is given in [64]   □

Based on **Theorem** 5.1, we can reduce the IR constraints of problem **P1** from M to 1, and the IC constraints from $M(M-1)$ to $M-1$ [64, 65]. More details can be found in the proof of **Theorem** 5.1.

To further simplify the constraints, we give the following theorem.

Theorem 5.2 *The contract item designed for type 1 device, i.e., (T_1, B_1), must satisfy*

$$U_1(T_1, B_1) = \theta_1 B_1 - \gamma T_1 = 0. \tag{5.5}$$

The optimal contract item designed for any type m device, i.e., (T_m, B_m), $m \in M$, $m > 1$, must satisfy

$$\gamma T_m = \gamma T_{m-1} + \theta_m(B_m - B_{m-1}). \tag{5.6}$$

Proof For type 1 device, the BS can continuously increase T_1 or decrease B_1 to maximize its utility until $U_1(T_1, B_1) = 0$. For any type m device, $m \in M$, $m \ne 1$, we have

$$\gamma T_m \le \gamma T_{m-1} + \theta_m(B_m - B_{m-1}). \tag{5.7}$$

Thus, the BS can decrease B_m or increase T_m to further improve its own utility until the equality holds.   □

Based on the above analysis, **P1** can be rewritten as

$$\mathbf{P2}: \max_{\{T_m\}, \{B_m\}} U_{BS}(\{T_m\}, \{B_m\}),$$

$$\text{s.t. } C_1: \theta_1 B_1 - \gamma T_1 = 0,$$

$$C_2: \gamma T_m = \gamma T_{m-1} + \theta_m(B_m - B_{m-1}), 2 \le m \le M,$$

$$C_3, C_4, \forall m. \tag{5.8}$$

By checking the Hessian matrix, it can be verified that **P2** is a convex programming problem, which can be solved by using the KKT conditions. The detailed solution process is omitted here due to space limitation.

5.3   Resource Allocation Base on Lyapunov Optimization and Matching Theory

In this section, we introduce the long-term cross-layer online resource allocation mechanism based on Lyapunov optimization and matching theory.

5.3.1 Dynamic Queue Model

After access control, only the K delay-sensitive MTC devices, i.e., $\mathcal{DS}_K$, are connected to the BS. For device DS_k, the backlog Q_k of queue k evolves as

$$Q_k(t+1) = [Q_k(t) - v_k(t)]^+ + R_k(t), \qquad (5.9)$$

where $[\cdot]^+$ represents $\max[\cdot, 0]$. Q_k is mean rate stable [66] if

$$\lim_{T \to \infty} \frac{\mathbb{E}\{|Q_k(T)|\}}{T} = 0. \qquad (5.10)$$

We define D_k as the transmission delay for queue k. By *Little' Law*, the average delay constraint is given by

$$D_k = \frac{\lim\limits_{T \to \infty} \frac{1}{T} \sum_{t=0}^{T-1} \mathbb{E}\{Q_k(t)\}}{\lim\limits_{T \to \infty} \frac{1}{T} \sum_{t=0}^{T-1} \mathbb{E}\{v_k(t)\}} \leq D_{k,max}^Q, \qquad (5.11)$$

where $D_{k,max}^Q$ is the maximum tolerable transmission delay.

For device DS_k, the application-layer satisfaction U_k is positively related to the sensing rate $R_k(t)$. For example, the supervisory control and data acquisition (SCADA) service in power system automation requires at least a two-second measurement and refreshment rate [67]. U_k can be defined as

$$U_k[R_k(t)] = \alpha_k \log_2[R_k(t)], \qquad (5.12)$$

where α_k is a service weight parameter which indicates the importance or priority of $R_k(t)$ to DS_k. Intuitively, device DS_k with a larger α_k should be allocated with a higher sensing rate R_k because a larger α_k also represents that U_k increases significantly as $R_k(t)$ increases. The logarithmic function represents that the marginal increment of satisfaction declines gradually with $R_k(t)$.

We assume that there are N orthogonal subchannels, the set of which is defined as $S_N = \{S_1, \cdots, S_n, \cdots, S_N\}$, where $n \in \mathcal{N} = \{1, \cdots, n, \cdots, N\}$. The channel selection indicator is denoted as $x_{k,n}(t)$, which is a binary value. $x_{k,n}(t) = 1$ means that subchannel S_n is allocated to device DS_k, and otherwise, $x_{k,n}(t) = 0$. The transmission rate of device DS_k is given by

$$v_k(t) = \sum_{n=1}^{N} x_{k,n}(t) W_n(t) \log_2 \left(1 + \frac{p_k(t) g_{k,n}(t)}{\sigma_0}\right), \tag{5.13}$$

where $W_n(t)$ denotes the bandwidth of subchannel S_n, and $p_k(t)$ is the transmission power of device DS_k. $g_{k,n}(t)$ is the channel gain. σ_0 is the noise power.

In reality, the lifetime and the connectivity of a M2M network are highly dependent on the battery states of underlying MTC devices. Once a MTC device has been deployed, it is difficult to replace its battery. Therefore, a long-term average power consumption constraint is needed to ensure the reliable operation of the M2M network, which is given by

$$0 \le \lim_{T \to \infty} \frac{1}{T} \sum_{t=0}^{T-1} \mathbb{E}\{E(t)\} \le P_{mean}, \tag{5.14}$$

where $E(t) = \sum_{k=1}^{K} \sum_{n=1}^{N} x_{k,n}(t) p_k(t)$ is the total energy consumption. P_{mean} is the time-average constraint of power consumption.

5.3.2   Problem Formulation and Transformation

The objective is to maximize the time-average satisfaction of all the MTC devices, which is given by

$$\bar{U}(T) = \lim_{T \to \infty} \frac{1}{T} \sum_{t=0}^{T-1} \mathbb{E}\left\{ \sum_{k=1}^{K} U_k[R_k(t)] \right\}. \tag{5.15}$$

The stochastic optimization problem is formulated as

$$\mathbf{P3}: \max_{\{R_k(t)\},\{x_{k,n}(t)\},\{p_k(t)\}} \bar{U}(T),$$

s.t. $C_5 : 0 \le \sum_{k=1}^{K} p_k(t) \le P_{max},$

$C_6 : 0 \le \sum_{k=1}^{K} R_k(t) \le R_{max},$

$C_7 : x_{k,n}(t) \in \{0, 1\}, \ \forall k, n,$

$C_8 : \sum_{k=1}^{K} x_{k,n}(t) \le 1, \ \forall n,$

$C_9 : \sum_{n=1}^{N} x_{k,n}(t) \le 1, \ \forall k,$

$$C_{10} : \lim_{T \to \infty} \frac{1}{T} \sum_{t=0}^{T-1} \mathbb{E}\{E(t)\} \le P_{mean}, \ \forall k, n,$$

$$C_{11} : D_k \le D_{k,max}^{Q}, \ \forall k,$$

$$C_{12} : Q_k \text{ is mean rate stable}, \ \forall k, \tag{5.16}$$

where C_5 is the short-term constraint of transmission power. C_6 is the constraint of sensing rate. $C_7 \sim C_9$ are the channel selection constraints, i.e., one subchannel can be allocated to at most one device, and each device is allowed with at most one subchannel. C_{10} and C_{11} are the long-term constraints of transmission power and transmission delay. C_{12} is the queue stability constraint defined in (5.10). Lyapunov optimization is a powerful methodology for addressing long-term optimization problems, which requires less priori information and owns a low computational complexity compared with traditional methods, such as stochastic optimization and dynamic programming [68]. It has been widely utilized in resource allocation optimization. In [69], Bao et al. used Lyapunov optimization to derive the optimal resource allocation for a downlink nonorthogonal multiple access (NOMA) system. In [70], Guo et al. introduced a Lyapunov-based cross-layer joint rate control and resource allocation scheme to maximize the time-average users' satisfaction. In [71], an online resource allocation algorithm based on Lyapunov optimization was introduced to address the energy efficiency maximization problem in multimedia heterogeneous cloud radio access networks (H-CRANs), which takes into account both the individual front-haul capacity and co-channel interference constraints.

One advantage of Lyapunov optimization is to transform a long-term optimization problem into a series of short-term subproblems, and transform the long-term optimization constraints into constraints with queue stability [66, 73, 74]. The implementation process is described as follows.

The long-term time-average constraints of **P3** can be transformed to queue stability constraints by exploiting the concept of virtual queue [69].

The virtual queues associated with constraints C_{10} and C_{11} are given by

$$Z(t + 1) = [Z(t) - P_{mean}]^+ + E(t),$$

$$Y_k(t + 1) = [Y_k(t) - v_k(t)D_{k,max}^{Q}]^+ + Q_k(t). \tag{5.17}$$

Theorem 5.3 *If $Z(t)$ and $Y_k(t)$ are mean rate stable, C_{10} and C_{11} hold automatically.*

Proof The details are omitted here due to space limitation. A similar proof can be found in [72]. □

Based on the **Theorem** 5.3, **P3** can be rewritten as

$$\mathbf{P4} : \max_{\{R_k(t)\}, \{x_{k,n}(t)\}, \{p_k(t)\}} \bar{U}(T),$$

s.t. $C_5 \sim C_9$,

$C_{10} : Q_k$, Y_k and Z are mean rate stable, $\forall k$. (5.18)

5.3.3 Joint Rate Control, Power Allocation and Channel Selection

Let $\mathbf{G}(t) = [Q(t), Y(t), Z(t)]$ be the concatenated vector of the data queue and virtual queues. The Lyapunov function is defined as [74]

$$L(\mathbf{G}(t)) = \frac{1}{2}\sum_{k=1}^{K} Q_k^2(t) + \frac{1}{2}\sum_{k=1}^{K} Y_k^2(t) + \frac{1}{2}Z^2(t). \qquad (5.19)$$

The Lyapunov drift is defined as the conditionally expected change of the Lyapunov function in two consecutive time slots. The one-step conditional Lyapunov drift is given by

$$\Delta(\mathbf{G}(t)) \triangleq \mathbb{E}\{L(\mathbf{G}(t+1)) - L(\mathbf{G}(t))|\mathbf{G}(t)\}. \qquad (5.20)$$

Intuitively, a smaller Lyapunov drift is essential to guarantee the queue stability. More details can be found in [68].

To maximize $\bar{U}(T)$ under the constraint of queue stability, we define the drift-minus-reward term as

$$DM(\mathbf{G}(t)) = \Delta(\mathbf{G}(t)) - V\mathbb{E}\{U(t)|\mathbf{G}(t)\}, \qquad (5.21)$$

where $U(t) = \sum_{k=1}^{K} U_k[R_k(t)]$. V is a nonnegative weight parameter which represents the relative importance of the drift $\Delta(\mathbf{G}(t))$ compared with the reward $\mathbb{E}\{U(t)|\mathbf{G}(t)\}$, i.e., the tradeoff between "queue stability" and the "reward maximization".

Theorem 5.4 *Under all possible $\mathbf{G}(t)$ and $V \geq 0$, the upper bound of the drift-minus-reward term is given by*

$$\Delta(\mathbf{G}(t)) - V\mathbb{E}\{U_k(t)|\mathbf{G}(t)\}$$

$$\leq \sum_{k=1}^{K} \mathbb{E}\{Q_k(t)R_k(t) - VU_k[R_k(t)]|\mathbf{G}(t)\}$$

$$+ \sum_{k=1}^{K} Y_k(t)\mathbb{E}\{Q_k(t) - v_k(t)D_{k,max}^{Q}|\mathbf{G}(t)\}$$

$$+ Z(t)\mathbb{E}\{E(t) - P_{mean}|\mathbf{G}(t)\}$$

$$- \sum_{k=1}^{K} Q_k(t)\mathbb{E}\{v_k(t)|\mathbf{G}(t)\} + \Phi, \qquad (5.22)$$

where Φ is a positive constant that satisfies the following constraint:

$$\Phi \geq \frac{1}{2} \sum_{k=1}^{K} \mathbb{E}\{R_k^2(t) + v_k^2(t)|\mathbf{G}(t)\}$$

$$+ \frac{1}{2} \sum_{k=1}^{K} \mathbb{E}\{Q_k^2(t) + v_k^2(t)D_{k,max}^{Q}{}^2|\mathbf{G}(t)\}$$

$$+ \frac{1}{2}\mathbb{E}\{E^2(t) + P_{mean}^2|\mathbf{G}(t)\}. \tag{5.23}$$

Proof The detailed proof is omitted here due to space limitation. A similar proof can be found in [69]. □

Based on the principle of Lyapunov optimization, we can optimize the upper bound of the drift-minus-reward term defined in (5.22) at each time slot t subject to constraints $C_5 \sim C_9$.

It is noted that the first term of the right-hand side of (5.22) involves only the rate control variables $\{R_k(t)\}$, which can be expressed as

$$\sum_{k=1}^{K} f_1[R_k(t)], \tag{5.24}$$

where

$$f_1[R_k(t)] = Q_k(t)R_k(t) - VU_k[R_k(t)]. \tag{5.25}$$

The second, third, and fourth terms of the right-hand side of (5.22) involve only the power allocation and channel selection variables $\{p_k(t)\}$ and $\{x_{k,n}(t)\}$, which can be written as

$$\sum_{k=1}^{K} f_2[p_k(t), x_{k,n}(t)] + Z(t)P_{mean} - \sum_{k=1}^{K} Y_k(t)Q_k(t), \tag{5.26}$$

where

$$f_2[p_k(t), x_{k,n}(t)] = [Q_k(t) + Y_k(t)D_{k,max}^{Q}]v_k(t) - Z(t)p_k(t). \tag{5.27}$$

Thus, **P4** can be decoupled into a rate control subproblem, as well as a joint power allocation and channel selection subproblem, which are independent of each other.

Rate Control

The rate control subproblem is formulated as

$$\mathbf{P5} : \min_{\{R_k(t)\}} \sum_{k=1}^{K} f_1[R_k(t)],$$

$$\text{s.t.} \quad C_6. \tag{5.28}$$

P5 is a convex programming problem and can be solved by using the KKT conditions. The detailed solution process is omitted here due to space limitation.

Joint Power Allocation and Channel Selection

$Q_k(t)$, $Y_k(t)$ and $Z(t)$ represent the queue backlog values, which can be regarded as deterministic variables in each time slot. P_{mean} is the time-average constraint of power consumption, which is set as a constant. To simplify the problem, we can remove the term $Z(t)P_{mean} - \sum_{k=1}^{K} Y_k(t)Q_k(t)$ from (5.26), which is a constant that does not contain the optimization variables $\{p_k(t)\}$ and $\{x_{k,n}(t)\}$. The joint power allocation and channel selection problem can be written as

$$\textbf{P6}: \max_{\{p_k(t)\},\{x_{k,n}(t)\}} \sum_{k=1}^{K} f_2[p_k(t), x_{k,n}(t)],$$

$$\text{s.t.} \quad C_5, C_7 \sim C_9. \tag{5.29}$$

P6 is NP-hard since the integer variables and the continuous variables are coupled together. To provide a tractable solution, **P6** can be transformed into a two-dimensional matching problem, which is described as a triple $(\mathcal{DS}_K, \mathcal{S}_N, \mathcal{F})$. Here, $\mathcal{DS}_K$ and $\mathcal{S}_N$ represent the sets of matching participants, i.e., delay-sensitive MTC devices and subchannels, respectively. $\mathcal{F}$ represents the set of devices' preferences. A matching ϕ is defined as:

Definition 5.4 (Matching) ϕ is a one-to-one correspondence from set $\mathcal{DS}_K \cup \mathcal{S}_N$ onto itself under preference $\mathcal{F}$, i.e., $\phi(DS_k) \in \mathcal{S}_N$, $\forall k \in \mathcal{K}$. $\phi(DS_k) = S_n$ represents that device DS_k is matched with subchannel S_n, i.e., $x_{k,n}(t) = 1$.

When $\phi(DS_k) = S_n$, the maximum value of $f_2[p_k(t) \mid x_{k,n}(t) = 1]$ can be obtained by solving the following power allocation problem:

$$\textbf{P7}: \max_{p_k(t)} f_2[p_k(t) \mid x_{k,n}(t) = 1],$$

$$\text{s.t.} \quad C_5. \tag{5.30}$$

P7 is also a convex optimization problem and can be solved by applying KKT conditions. The Lagrangian associated with **P7** is given by

$$\mathcal{L}(\{p_k(t)\}, \mu) \tag{5.31}$$

$$= -f_2[p_k(t) \mid x_{k,n}(t) = 1] + \mu(\sum_{k=1}^{K} p_k - p_{max}), \tag{5.32}$$

where μ is the Lagrange multiplier corresponding to constraint C_5. The optimal solution $p_k^*(t)$ is given as

$$p_k^*(t) = \min\{\frac{\ln 2C_k(t)}{Z(t)} - \frac{\sigma_0}{g_{k,n}(t)},$$

$$\frac{C_k(t)}{C(t)}\mathcal{P}_{max} + \sigma_0\frac{C_k(t)}{C(t)g(t)} - \frac{\sigma_0}{g_{k,n}(t)}\}, \quad (5.33)$$

where $C_k(t) = Q_k(t) + Y_k(t)D_{k,max}^Q$, $C(t) = \sum_{k=1}^K C_k(t)$ and $\frac{1}{g(t)} = \sum_{k=1}^K \frac{1}{g_{k,n}(t)}$. From (5.33), we can notice that $p_k^*(t)$ is positively related to $Q_k(t)$ and $Y_k(t)$, and is negatively related to $g_{k,n}(t)$. The detailed solution process is omitted here due to space limitation.

We define the preference of device DS_k towards subchannel S_n as

$$F_{DS_k,S_n}\big|_{\phi(DS_k)=S_n} = f_2[p_k^*(t) \mid x_{k,n}(t) = 1] - \Lambda_n, \quad (5.34)$$

where Λ_n is the virtual price of subchannel S_n, which is added to resolve the conflict of matching. It can be set as zero initially.

Thus, by temporally matching device DS_k with every subchannel, we can obtain its preference towards all the subchannels. The preference list of DS_k, i.e., $\mathcal{F}_k$, is constructed by sorting all the N subchannels in a descending order according to the preferences, i.e., $F_{DS_k,S_n}\big|_{\phi(DS_k)=S_n}, \forall S_n \in S_N$. The total set $\mathcal{F}$ is constructed as $\mathcal{F} = \{\mathcal{F}_k, \forall k \in \mathcal{K}\}$. Then the introduced pricing-based stable matching approach is implemented as follows.

Initially, set $\phi = \emptyset$, $\Omega = \emptyset$, and $\Lambda_n = 0$, $\forall n \in N$. Ω represents the set of subchannels which receive more than one matching proposal from MTC devices.

In the proposal process, if $\exists \phi(DS_k) = \emptyset$, device $DS_k \in DS_K$ will introduce to the subchannel which ranks as the first place in $\mathcal{F}_k$. Afterwards, if any subchannel $S_n \in S_N$ receives only one proposal from a device, then they will be directly matched. Otherwise, if S_n receives more than one proposal, add S_n into set Ω, and enter into the price rising process.

In the price pricing process, any subchannel $S_n \in \Omega$ increases its price Λ_n by $\Delta\Lambda_n$. Accordingly, all the devices that are competing for S_n update their preferences towards S_n and renew their proposals. A device may give up S_n if its matching cost is too high. The price rising process will continue until only one device remains, which is matched with S_n. Then, S_n is removed from Ω.

The matching will be finished until every device has been matched under the assumption of $N \geq K$.

5.4 Performance Results and Discussions

In this section, the introduced scheme is validated via simulations. The details are given as follow.

5.4.1 Feasibility and Efficiency of Access Control Mechanism

First, we verify the feasibility and efficiency of the introduced contract-based incentive mechanism for access control. We consider a single cell with $J = 20$ delay-tolerant MTC devices which are classified into $M = 20$ types, i.e., a device per type. ACB barring factor B_0 is 0.5. The probability of MTC device type P_m is assumed to follow a uniform distribution within the range $[0, 1]$. The profit coefficient of BS γ_0 is set to 0.5. The unit costs of BS β_0 and devices γ are set to 0.1 and 0.2, respectively.

Figure 5.2a and b show the waiting time and access probability versus different types of delay-tolerant MTC devices, respectively. Both the waiting time and the access probability increase monotonically with the device type, which is consistent with Theorem 5.1 and satisfies the monotonicity constraint.

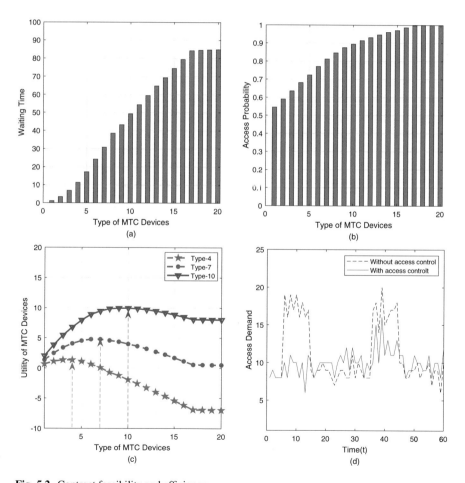

Fig. 5.2 Contract feasibility and efficiency

Figure 5.2c describes the utilities of type 4, type 7, and type 10 devices versus different types of contract items. It is observed that the contract is incentive compatible, i.e., a device achieves the maximum utility if and only if it selects the contract item designed for its own type. Thus, the contract can elicit the hidden information of device type.

Figure 5.2d shows the access control performance. We consider a total period of 60 s. The arrived number of delay-sensitive MTC devices per second follows a normal distribution with mean 8 and variance 1. The time periods [6, 15] and [36, 45] are defined as the peak time, during which a number of delay-tolerant MTC devices access to the BS simultaneously. The number of delay-tolerant MTC devices per second follows the same normal distribution. Simulation results show that the peak-time access demand can be effectively flattened after applying access control. The rationale behind is that some delay-tolerant MTC devices postpone their access to exchange for a higher access probability, which effectively shift the access demands from the peak time to the non-peak time.

5.4.2   Feasibility and Efficiency of Resource Allocation Scheme

In this subsection, we evaluate the resource allocation scheme for a total of $T = 100$ time slots. At each time slot, the joint optimization of rate control, power allocation, and channel selection is carried out for $K = 4$ delay-sensitive MTC devices, i.e., DS_1 to DS_4. The number of subchannels N is also 4. The bandwidth of each subchannel is 20 MHz. P_{mean} and P_{max} are set as 0.8 W and 1 W, respectively. The upper bound of sum sensing rates R_{max} is set as 20 Mbits/s. $D_{1,max}^Q$ to $D_{4,max}^Q$ are set as 0.2, 0.4, 0.6, and 0.8 s, respectively. α_1 to α_4 are set as 0.1, 0.2, 0.3, and 0.4, respectively. The increment of price $\Delta\Lambda$ is set as 0.1. We set $V = 100$.

Figure 5.3a–c show the evolution of queue backlogs Q_k, Z and Y_k over time, respectively. It is observed that queue backlogs tend to be stable within a short period of time, which guarantees the stability of queues. Figure 5.4d shows the sensing rate versus time slot. We notice that both Q_k and Y_k are positively related to α_k. The reason is that device DS_k with a larger α_k is generally allocated with a higher R_k to maximize the application-layer satisfaction U_k. Based on (5.9), a higher $R_k(t)$ also results in a larger data queue backlog Q_k due to the large amount of data entering the queue. Y_k is the virtual queue associated with the delay constraint. Since Y_k is positively related to Q_k based on (5.17) while Q_k is positively related to α_k, we can derive that Y_k is also positively related to α_k.

Figure 5.4a–d show the transmission rate and delay versus time slot. Simulation results show that the transmission rates of the introduced algorithm are dynamically optimized in accordance with the sensing rates, while the baseline algorithm optimizes the transmission rates without awareness of the sensing rates. Thus, DS_1 with higher sensing rate will experience severe performance degradation in terms of delay as shown in Fig. 5.4d. We notice that there exists a large fluctuation of the transmission rate. The transmission rate is mainly affected by physical-layer parameters, such as subchannel bandwidth, transmission power, and channel

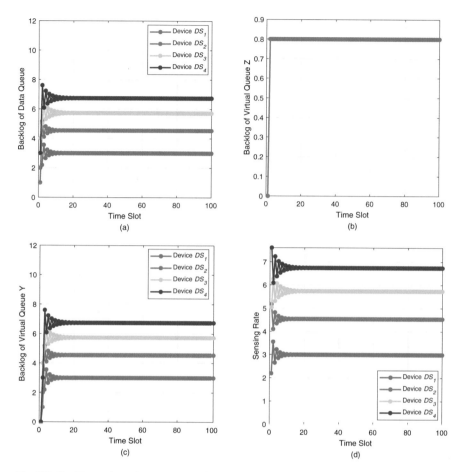

Fig. 5.3 Backlog and sensing rate versus time slot

conditions, etc. In simulations, the channel gain is varied randomly in each slot, which causes a large fluctuation of the transmission rate. However, despite the time-varying channel gains and the large fluctuation of the transmission rate, both the optimality and queue stability are guaranteed.

Furthermore, we can notice that there exists a big gap between the sensing rate and the transmission rate in Figs. 5.3d and 5.4a. The reason is that if Q_k and Y_k are mean rate stable, we can obtain $v_k(t) \geq R_k(t) = Q_k(t)$ and $v_k(t) \geq Q_k(t)/D^Q_{k,max}$ based on (5.9) and (5.17), respectively. For device DS_k, the maximum delay tolerance $D^Q_{k,max}$ is usually less than the slot duration ΔT, i.e., $D^Q_{k,max} < \Delta T \leq 1$, which will cause that $\dfrac{Q_k(t)}{D^Q_{k,max}}$ is much larger than $R_k(t)$. Thus, we have $v_k(t) \geq \dfrac{Q_k(t)}{D^Q_{k,max}} >> R_k(t) = Q_k(t)$, which indicates that there exists a big gap between the sensing rate and transmission rate.

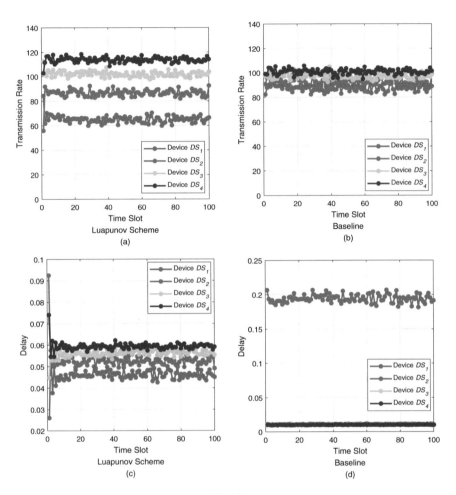

Fig. 5.4 Transmission rate and delay versus time slot

In this chapter, we introduced a strategy jointly considering access control and resource allocation for M2M communications in industrial automation. First, we developed a contract-based incentive mechanism to motivate delay-tolerant devices to postpone their access voluntarily. Then, a joint rate control, power allocation and channel selection problem was formulated with the special consideration including long-term constraints of delay and power consumption.

Numerical results demonstrated that the introduced scheme can effectively reveal the MTC device type under information asymmetry, and achieve a good performance in flattening the peak-time traffic demand. It also achieves significant performance improvement in terms of sensing rate, queue stability, backlog fluctuation, and energy efficiency compared with the snapshot-based throughput optimal algorithm.

Chapter 6
Energy-Efficient Context-Aware Resource Allocation for Edge-Computing-Empowered Industrial IoT

6.1 Framework of Energy-Efficient Edge-Computing-Empowered IIoT

In this section, the system model and problem formulation are introduced.

6.1.1 System Model

As shown in Fig. 6.1, we consider a single-cell scenario where an edge server is collocated with a BS. The BS provides connection service and the edge server provides computing service for K MTDs within the cell, the set of which is denoted by $\mathcal{M} = \{m_1, \cdots, m_k, \cdots, m_K\}$. There exist J orthogonal subchannels, the set of which is defined as $C = \{c_1, \cdots, c_j, \cdots, c_J\}$. The bandwidth of subchannel c_j is denoted by B_j. Channel selection conflict occurs when more than one MTDs select the same subchannel at the same time, and only one MTD can succeed to access the subchannel under the coordination of the BS.

A time-slotted model is adopted where the total optimization period is divided into T slots with equal length τ, the set of which is denoted by $\mathcal{T} = \{1, \cdots, t, \cdots, T\}$. In this model, CSI remains unchanged within a slot and varies across different slots. In each slot, each MTD determines its channel selection strategy individually. Particularly, a MTD faces $J + 1$ options, i.e., either selecting one of the J subchannels or remaining idle. Figure 6.1 shows an example of channel selection with 4 MTDs and 2 subchannels. m_1 selects subchannel c_1 for data transmission while m_2 remains idle. Channel selection conflict occurs between m_3 and m_4 due to the simultaneous selection of subchannel c_2.

In the following, the models of task transmission, energy consumption, delay, and service reliability are introduced.

© Springer Nature Switzerland AG 2021
Z. Zhou et al., *Green Internet of Things (IoT): Energy Efficiency Perspective*,
Wireless Networks, https://doi.org/10.1007/978-3-030-64054-5_6

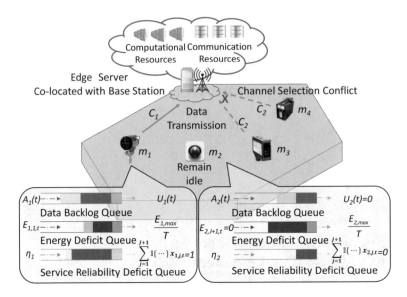

Fig. 6.1 Framework of edge-computing-empowered IIoT

Task Transmission Model

In the t-th slot, $A_k(t)$ new tasks with equal size γ_k arrive at $m_k \in \mathcal{M}$, which are firstly stored in the local buffer and then are transmitted to the edge server. Hence, the total task size is $\gamma_k A_k(t)$. Meanwhile, it has to retransmit $Y_k(t)$ amount of data, which have not been correctly delivered due to bit error. The task data stored in the local buffer of m_k can be modeled as a queue, i.e., queue k. $\gamma_k A_k(t)$ as well as $Y_k(t)$ can be seen as the amount of task data entering the queue and $U_k(t)$ represents the amount of task data leaving the queue. Define $Q_k(1)$ as the initial amount of data backlog. $Q_k(t)$ is the backlog of data queue k in the t-th slot, i.e., an accumulation of data that are yet to be processed. $Q_k(t)$ is dynamically evolved as

$$Q_k(t+1) = \max\{Q_k(t) - U_k(t), 0\} + \gamma_k A_k(t) + Y_k(t+1). \tag{6.1}$$

The set of channel selection indicators consists of $J+1$ binary elements, which is denoted by $\{x_{k,j,t}\}$, where $x_{k,j,t} \in \{0, 1\}$. When $j = 1, 2, \cdots, J$, $x_{k,j,t} = 1$ represents that m_k selects subchannel c_j for data transmission in the t-th slot and when $j = J+1$, $x_{k,j,t} = 1$ represents that m_k remains idle.

Considering the powerful computational capability of the edge server, the objective of each MTD is to offload as many tasks as possible, which equals to maximizing the total amount of task data that can be transmitted, i.e., the throughput.

Uplink transmission is considered here. Denote $H_{k,j,t}$ as the uplink channel gain of subchannel c_j between m_k and the BS. Given $x_{k,j,t}$, the achievable uplink transmission rate is given by

$$R_{k,j,t} = \begin{cases} B_j \log_2(1 + \frac{P_{TX} H_{k,j,t}}{\delta^2}), & j = 1, 2, \cdots, J \\ 0, & j = J + 1 \end{cases}, \tag{6.2}$$

where δ^2 is the noise power, and P_{TX} is the transmission power. The throughput of m_k in the t-th slot is given by

$$z_{k,j,t} = \min\{Q_k(t), \tau R_{k,j,t}\}. \tag{6.3}$$

The amount of data transmitted to the edge server can be

$$U_k(t) = \sum_{j=1}^{J+1} x_{k,j,t} z_{k,j,t}. \tag{6.4}$$

Denote the bit error rate (BER) for m_k transmitting data through subchannel c_j in the t-th slot as $P_{k,j,t}^e$. We consider the noncoherent binary phase shift keying (BPSK) modulation and the corresponding BER [75] of it can be derived as

$$P_{k,j,t}^e = \frac{1}{2} \text{erfc}\left(\sqrt{\frac{P_{TX} H_{k,j,t}}{\delta^2}}\right). \tag{6.5}$$

Here, BPSK is just used as an example to derive the queue evolution model, which can be naturally extended to other modulation schemes such as quadrature amplitude modulation (QAM) and orthogonal frequency division multiplexing (OFDM).

Therefore, $Y_k(t + 1)$, the amount of data that has to be retransmitted in the next slot can be calculated as

$$Y_k(t + 1) = U_k(t) P_{k,j,t}^e. \tag{6.6}$$

Energy Consumption Model

In the t-th slot, the energy consumption of m_k for data transmission is the transmission power multiplied by the transmission delay, i.e.,

$$E_{k,j,t} = \begin{cases} P_{TX} \min\{\frac{Q_k(t)}{R_{k,j,t}}, \tau\}, & j = 1, 2, \cdots, J. \\ 0, & j = J + 1. \end{cases} \tag{6.7}$$

The limited battery capacity exerts a direct impact on the total energy budget of m_k over T slots, which is denoted by $E_{k,max}$. Therefore, the long-term energy consumption of m_k must satisfy

$$E_k = \sum_{t=1}^{T} \sum_{j=1}^{J+1} x_{k,j,t} E_{k,j,t} \leq E_{k,max}. \tag{6.8}$$

Delay Model

In IIoT, the data size of computational results is generally smaller than that of the computational tasks. Therefore, for the sake of simplicity, we can neglect the downlink transmission delay. Some previous works, e.g., [76–78], also ignore the downlink transmission time. On the other hand, our work can be easily extended to the scenario where the downlink transmission time is considered. Therefore, the total offloading delay is the sum of the transmission delay and computational delay, which can be given by

$$d_{k,j,t}^{total} = d_{k,j,t}^{tra} + d_{k,j,t}^{com}. \tag{6.9}$$

Given $x_{k,j,t}$ and $z_{k,j,t}$, the transmission delay is calculated by dividing throughput $z_{k,j,t}$ with transmission rate $R_{k,j,t}$, i.e.,

$$d_{k,j,t}^{tra} = \begin{cases} \frac{z_{k,j,t}}{R_{k,j,t}} = \min\{\frac{Q_k(t)}{R_{k,j,t}}, \tau\}, & j = 1, 2, \cdots, J. \\ +\infty, & j = J+1. \end{cases} \tag{6.10}$$

Based on the computational intensity model in [79], assuming that the computational intensity of the task data transmitted by m_k in the t-th slot is $\lambda_{k,t}$ (CPU cycles/bit), it requires $z_{k,j,t}\lambda_{k,t}$ CPU cycles to process the task data. It is noted that although a linear relationship between workload and data size is employed, our work is compatible with other nonlinear models and can be used for different kinds of IIoT applications with different computing intensities. Denoting the available computational resources for m_k in the t-th slot as $\xi_{k,t}$, the computational delay is calculated as

$$d_{k,j,t}^{com} = \begin{cases} \frac{z_{k,j,t}\lambda_{k,t}}{\xi_{k,t}}, & j = 1, 2, \cdots, J. \\ +\infty, & j = J+1. \end{cases} \tag{6.11}$$

Service Reliability Requirement Model

We model the service reliability requirement in terms of delay. Denoting the task delay requirement as $d_{k,t}$, the task offloading is unsuccessful if the offloaded task cannot be processed within the specified delay requirement, i.e., $d_{k,j,t}^{total} > d_{k,t}$. Denote $X_{k,T}$ as the number of successful task offloading for m_k over T slots, which is given by

$$X_{k,T} = \sum_{t=1}^{T} \sum_{j=1}^{J+1} \mathbb{I}\{d_{k,j,t}^{total} \le d_{k,t}\} x_{k,j,t}. \tag{6.12}$$

$\mathbb{I}\{x\}$ is an indicator function with $\mathbb{I}\{x\} = 1$ if event x is true and $\mathbb{I}\{x\} = 0$ otherwise. The edge server performs computational resource optimization at the end of each slot and feeds back the result of whether the delay requirement of m_k can be satisfied or not.

The service reliability requirement is defined as

$$\frac{X_{k,T}}{T} \ge \eta_k, \tag{6.13}$$

where $\eta_k \in (0, 1]$ represents the minimum successful probability of task offloading.

6.1.2 Problem Formulation

The objective is to maximize the long-term network throughput under the long-term constraints of energy budget and service reliability. Therefore, network throughput maximization problem is formulated as

$$\textbf{P1}: \max_{\{x_{k,j,t}\}} \sum_{t=1}^{T} \sum_{k=1}^{K} \sum_{j=1}^{J+1} x_{k,j,t} z_{k,j,t},$$

$$\text{s.t.}\ \ C_1: \sum_{k=1}^{K} x_{k,j,t} \le 1, j = 1, 2, \cdots, J, \forall t \in \mathcal{T},$$

$$C_2: \sum_{j=1}^{J+1} x_{k,j,t} \le 1, \forall m_k \in \mathcal{M}, \forall t \in \mathcal{T},$$

$$C_3: \sum_{t=1}^{T} \sum_{j=1}^{J+1} x_{k,j,t} E_{k,j,t} \le E_{k,max}, \forall m_k \in \mathcal{M},$$

$$C_4: \frac{X_{k,T}}{T} \ge \eta_k, \forall m_k \in \mathcal{M}, \tag{6.14}$$

where C_1 and C_2 are the channel selection constraints, i.e., in each slot, each subchannel can be selected by at most one MTD, and each MTD can select only one subchannel at most or remains idle. C_3 and C_4 correspond to the constraints of energy consumption and service reliability, respectively. Here, we focus on optimizing channel selection strategy while the optimization of computational resource allocation is left to the future work. The reason is that the algorithm is naturally compatible with any computational resource allocation scheme. Similarly, some previous works also only consider the channel selection problem [79–81]. On the other hand, the joint optimization of channel selection and computational resource allocation is a completely different problem, which requires different system modeling, problem formulation, and optimization design. Utilizing learning algorithms to solve the joint optimization problem of integer channel selection and continuous computational resource allocation is also a worthwhile research direction which will be investigated in the future work.

6.2 Learning-Based Context-Aware Channel Selection for the Single-MTD Scenario

In this section, we consider the single-MTD scenario with only one MTD, e.g., m_k, and propose a learning-based context-aware channel selection algorithm.

6.2.1 Lyapunov Based Problem Transformation

Problem **P1** cannot be directly solved due to the long-term optimization objective and constraints. To provide a tractable solution, we leverage Lyapunov optimization to transform a coupled long-term stochastic optimization problem into a series of short-term deterministic problems [68, 82], which can be solved in low complexity while the data backlog, energy consumption, and service reliability are balanced over time. Based on the concept of virtual queue [83], the long-term energy budget and service reliability constraints, i.e., C_3 and C_4, can be transformed to queue stability constraints. We define a virtual energy deficit queue $N_k(t)$ and a virtual service reliability deficit queue $F_k(t)$, which are evolved as

$$N_k(t+1) = \max\{N_k(t) + \sum_{j=1}^{J+1} x_{k,j,t} E_{k,j,t} - \frac{E_{k,max}}{T}, 0\},$$

$$F_k(t+1) = \max\{F_k(t) + \eta_k - \sum_{j=1}^{J+1} \mathbb{I}\{d_{k,j,t}^{total} \le d_{k,t}\} x_{k,j,t}, 0\}, \tag{6.15}$$

with $N_k(1) = F_k(1) = 0$. $N_k(t)$ represents the deviation of current energy consumption from the energy budget, while $F_k(t)$ reflects the deviation of service reliability from the specified requirement. Examples of queue evolution for MTDs are shown in Fig. 6.1. Taking m_1 as an example, the data queue $Q_1(t)$, the virtual energy deficit queue $N_1(t)$, and the virtual service reliability deficit queue $F_1(t)$ are dynamically updated at each slot based on (6.1) and (6.15). Comparing m_1 and m_2, it is noted that the data backlog and the service reliability deficit of m_1 are larger while the energy deficit of m_2 is larger.

Then, **P1** can be transformed into a series of short-term optimization subproblems. At each slot, if the energy consumption of m_k until the t-th slot does not exceed the energy budget, an online multi-objective optimization problem is defined to maximize throughput and service reliability while minimizing energy consumption, which is given by

$$\textbf{P2}: \min_{\{x_{k,j,t}\}} \sum_{k=1}^{K} \sum_{j=1}^{J+1} [-V_k z_{k,j,t} + \alpha_k N_k(t) E_{k,j,t}$$

$$- \beta_k F_k(t) (\sum_{j=1}^{J+1} \mathbb{I}\{d_{k,j,t}^{total} \leq d_{k,t}\} x_{k,j,t} - \eta_k)],$$

s.t. $C_1 \sim C_2$. (6.16)

For convenience, we write $\theta_{k,j,t} = -V_k z_{k,j,t} + \alpha_k N_k(t) E_{k,j,t} - \beta_k F_k(t)$ $(\sum_{j=1}^{J+1} \mathbb{I}\{d_{k,j,t}^{total} \leq d_{k,t}\} x_{k,j,t} - \eta_k)$. Here, $\theta_{k,j,t}$ is a weighted sum of throughput, energy consumption and service reliability, where V_k, $\alpha_k N_k(t)$, and $\beta_k F_k(t)$ are the corresponding weights.

P2 and **P1** are not equal, and the results of **P2** may not be feasible for **P1**. Furthermore, C_3 can be guaranteed by defining that if the energy budget of m_k is exhausted, then it cannot transmit data and is forced to remain idle. In other words, at the t-th slot, **P2** will be solved if and only if the energy budget is not exhausted. On the other hand, C_4 is satisfied in a best effort way due to service reliability awareness, i.e., a large deviation from the service reliability requirement will enforces m_k to select the option with higher successful chances of task offloading, thereby trying the best to satisfy C_4. It is noted that C_4 cannot be 100% guaranteed due to the lack of centralized optimization and coordination among all the MTDs.

The local information is referred as the information that can be possessed by m_k without additional information exchange with other entities in the network, e.g., the BS or the other MTDs. The nonlocal information refers to the information that can only be possessed by m_k with additional information exchange. Otherwise, if information exchange is infeasible, nonlocal information is unknown to m_k. Therefore, the information required to solve **P2** can be classified into two categories, i.e.,

- **Local information:** information that can be possessed by m_k without additional information exchange, e.g., the queue backlog $Q_k(t)$, the transmission power

---

Algorithm 6.1 SEB-GSI

1: Input: V_k, α_k, β_k.
2: **Phase 1:** Initialization
3: Set $Q_k(1)$ as the initial amount of data backlog, $N_k(1) = 0$, $F_k(1) = 0$, $x_{k,j,t} = 0$, $j = 1, 2, \cdots, J + 1, \forall t \in \mathcal{T}$.
4: **Repeat**
5: **Phase 2:** Decision making
6: Input: $H_{k,j,t}, \delta^2$.
7: Calculate the accurate value of $\theta_{k,j,t}$ with GSI, $j = 1, 2, \cdots, J + 1$.
8: Choose j by solving **P2**.
9: Observe $z_{k,j,t}$, $E_{k,j,t}$ and whether the delay requirement can be satisfied or not.
10: Update $U_k(t)$ and $Y_k(t + 1)$ based on (6.4) and (6.6).
11: Update $Q_k(t + 1)$, $N_k(t + 1)$, and $F_k(t + 1)$ as (6.1) and (6.15).
12: **Until** $t > T$.

P_{TX}, the total energy budget $E_{k,max}$, the computational intensity of task data $\lambda_{k,t}$, the task delay requirement $d_{k,t}$, and the service reliability requirement η_k.

- **Nonlocal information:** information that cannot be possessed by m_k without additional information exchange, e.g., the uplink channel gain $H_{k,j,t}$ for any subchannel $c_j \in C$, the available computational resources of the edge server $\xi_{k,t}$, and the channel selection strategies of other MTDs $\{x_{k,j,t}\}$ (only required for the multi-MTD scenario).

For the local information, the time-varying information is denoted by the symbol with subscript t or as a function of t, e.g., $Q_k(t)$, $\lambda_{k,t}$, and $d_{k,t}$. Otherwise, the local information is fixed, e.g., P_{TX}, $E_{k,max}$, and η_k. The nonlocal information is expressed in the same way and all the nonlocal information is time-varying.

Based on whether m_k has the nonlocal information or not, we consider an ideal and nonideal case, respectively. In the ideal case, m_k has the perfect knowledge of GSI, which includes both local and nonlocal information. In the nonideal case, m_k only knows the local information while the nonlocal information is unavailable.

6.2.2   SEB-GSI Algorithm for the Ideal Case

For the ideal case with GSI, we propose a service-reliability-aware, energy-aware, and data- backlog-aware GSI (SEB-GSI) algorithm for channel selection. SEB-GSI does not require future non-causal information. The detailed procedures are summarized in Algorithm 6.1, which consists of two phases, i.e., initialization (Line 2–3) and decision making (Line 5–9). Algorithm 6.1 is provided to demonstrate how to initialize queues, determine the optimal option, and update queues.

In the initialization phase, the initial length of all the queues and initial values of all the selection indicators are set as zero.

Then, the decision making phase is executed in a slot-by-slot fashion. At the beginning of the t-th slot, m_k calculates the value of $\theta_{k,j,t}$ towards option j,

---

Algorithm 6.2 SEB-UCB

1: Input: $V_k, \alpha_k, \beta_k, \omega$.
2: **Phase 1:** Initialization
3: Set $Q_k(1)$ as the initial amount of data backlog, $N_k(1) = 0$, $F_k(1) = 0$, $\bar{\theta}_{k,j,0} = 0$, $\hat{x}_{k,j,0} = 0$ and $x_{k,j,t} = 0$, $j = 1, 2, \cdots, J+1, \forall t \in \mathcal{T}$.
4: **Repeat**
5: **Phase 2:** Estimation and decision making
6: Calculate the estimation value of the MTD towards option j as (6.17).
7: Select the optimal option j based on (6.18).
8: **Phase 3:** Learning
9: Observe $z_{k,j,t}$, $E_{k,j,t}$ and whether the delay requirement can be satisfied or not.
10: Update $\bar{\theta}_{k,j,t}$ and $\hat{x}_{k,j,t}$ based on (6.19) and (6.20).
11: Update $U_k(t)$ and $Y_k(t+1)$ based on (6.4) and (6.6).
12: Update $Q_k(t+1)$, $N_k(t+1)$, and $F_k(t+1)$ as (6.1) and (6.15).
13: **Until** $t > T$.

$j = 1, 2, \cdots, J+1$, based on the current GSI. The optimum option j can be found by solving **P2**, which is equivalent to a minimum seeking problem with computational complexity $O(J)$. Afterwards, m_k sets $x_{k,j,t} = 1$, and updates all queues accordingly. In the next slot, the iteration continues until $t > T$.

The SEB-upper confidence bound (SEB-UCB) can adapt to the variations of the amount of data backlog, energy state and the service reliability state due to the endowed context awareness, which is achieved through the dynamic adjustment of channel selection strategy based on the values of $F_k(t)$, $N_k(t)$, and $Q_k(t)$. Details are given as follows:

- **Service reliability awareness:** When the service reliability deviates severely from the service reliability requirement, a large weight $F_k(t)$ will be placed on the service reliability term which enforces m_k to select the option with higher successful chances of task offloading, thereby enabling service reliability awareness.
- **Energy awareness:** When the energy consumption significantly exceeds the current energy budget, a large weight $N_k(t)$ on the energy consumption term will enforce m_k to select the option with less consumption, i.e., remaining idle, thereby enabling energy awareness.
- **Backlog awareness:** A large data backlog $Q_k(t)$ will lead to a large throughput $z_{k,j,t} = \tau R_{k,j,t}$ based on (6.3), which motivates m_k to choose the subchannel with higher data transmission rate, thereby enabling backlog awareness.

Since $F_k(t)$, $N_k(t)$, and $Q_k(t)$ are updated without requiring future information, SEB-GSI optimizes the balance among throughput performance, energy consumption, and service reliability requirement in an online fashion.

6.2.3 SEB-UCB Algorithm for the Nonideal Case

In the nonideal case where the nonlocal information is unavailable, the SEB-GSI algorithm is infeasible because the accurate value of $\theta_{k,j,t}$ cannot be obtained. To tackle this problem, we modify SEB-GSI based on the UCB1 framework [84], which is a low-complexity learning-based algorithm to deal with the sequential decision-making problem, and develop a learning-based context-aware channel selection algorithm named SEB-UCB. Instead of directly calculating $\theta_{k,j,t}$ in SEB-GSI, SEB-UCB estimates $\theta_{k,j,t}$ based on historical observations while simultaneously taking into account the uncertainty of estimation via confidence bound. It enables m_k to learn the optimal option based only on local information and achieve a bounded deviation from the optimal performance obtained with GSI.

The SEB-UCB algorithm is summarized in Algorithm 6.2. In each time slot, m_k makes decisions based on only two kinds of local information: $\bar{\theta}_{k,j,t-1}$ and $\hat{x}_{k,j,t-1}$, where $\bar{\theta}_{k,j,t-1}$ represents the empirical estimation of $\theta_{k,j,t-1}$ up to slot t, and $\hat{x}_{k,j,t-1}$ represents the number of times that m_k has selected the j-th option up to slot t. The estimation of m_k towards the option j in the t-th slot is estimated as

$$\widetilde{\theta}_{k,j,t} = \bar{\theta}_{k,j,t-1} - \omega \sqrt{\frac{2 \ln t}{\hat{x}_{k,j,t-1}}}, \tag{6.17}$$

where the first term represents the empirical performance of the option j, and the second term represents the confidence bound, which is designed to balance the tradeoff between exploration and exploitation. On one hand, the first term pushes m_k to select *a priori known optimal option* up to slot t. On the other hand, the second term is inversely proportional to $\hat{x}_{k,j,t-1}$, which allows m_k to explore options with less number of selections in order to improve the accuracy of estimation. Here, ω is the weight of exploration compared with exploitation, i.e., a larger ω represents a higher preference for exploration.

After estimating $\widetilde{\theta}_{k,j,t}$ for all the $J+1$ options, m_k chooses option j with the least estimation value, which is determined as

$$j = \arg\min_{j} \left\{ \widetilde{\theta}_{k,j,t} \right\}. \tag{6.18}$$

Then, m_k observes the corresponding results $z_{k,j,t}$, $E_{k,j,t}$ associated with $x_{k,j,t} = 1$ and whether the delay requirement can be satisfied or not. Accordingly, $\bar{\theta}_{k,j,t}$ and $\hat{x}_{k,j,t}$ are updated as

$$\bar{\theta}_{k,j,t} = \frac{\bar{\theta}_{k,j,t-1}\hat{x}_{k,j,t-1}}{\hat{x}_{k,j,t-1} + x_{k,j,t}}$$

$$+ \frac{-V_k z_{k,j,t} x_{k,j,t}}{\hat{x}_{k,j,t-1} + x_{k,j,t}} + \frac{\alpha_k N_k(t) E_{k,j,t} x_{k,j,t}}{\hat{x}_{k,j,t-1} + x_{k,j,t}}$$

$$-\frac{\beta_k F_k(t)(\sum_{j=1}^{J} \mathbb{I}\{d_{k,j,t}^{total} \le d_{k,t}\}x_{k,j,t} - \eta_k)x_{k,j,t}}{\hat{x}_{k,j,t-1} + x_{k,j,t}}, \qquad (6.19)$$

and

$$\hat{x}_{k,j,t} = \hat{x}_{k,j,t-1} + x_{k,j,t}. \qquad (6.20)$$

Based on $z_{k,j,t}$, $U_k(t)$ and $Y_k(t+1)$ can be calculated based on (6.4) and (6.6). Next, the three queues, i.e., $Q_k(t+1)$, $N_k(t+1)$, and $F_k(t+1)$, are updated as (6.1), (6.15).

Finally, increase t to $t+1$, and repeat lines 5–12 until $t > T$.

6.3 Learning-Based Context-Aware Channel Selection for the Multi-MTD Scenario

In this section, we consider channel selection under the multi-MTD scenario, where the channel selection strategies of different MTDs are coupled. Both the SEB-GSI and the SEB-UCB algorithms introduced in the previous section are not suitable for this scenario because the coupling among MTDs are not considered. To tackle this problem, we start from the ideal case with perfect GSI, and develop a matching-based context-aware channel selection algorithm named SEB-MGSI. Next, we consider the more practical nonideal case with only local information, and develop a matching-learning based context-aware channel selection algorithm named SEBC-MUCB.

6.3.1 SEB-MGSI Algorithm for the Ideal Case

When K MTDs are competing for the J subchannels, the channel selection problem involves a one-to-one matching between K MTDs and J subchannels. The definition of matching is given by

Definition 6.1 (Matching) Denote ϕ as the one-to-one correspondence from set $\mathcal{M} \cup \mathcal{C}$ onto itself. Specifically, $\phi(m_k) = c_j$ indicates that m_k is matched with subchannel c_j, i.e., $x_{k,j,t} = 1$, $j = 1, 2, \cdots, J$, and $\phi(m_k) = m_k$ indicates that m_k is not matched with any subchannel and has to remain idle, i.e., $x_{k,J+1,t} = 1$.

Remark 6.1 $x_{k,J+1,t} = 1$ actually contains two situations, the first of which is that m_k prefers to remain idle, and the second of which is m_k being forced to remain idle due to the shortage of subchannel.

The SEB-MGSI algorithm is developed based on pricing-based matching [59], which is summarized in Algorithm 6.3. It can be implemented in two phases:

Algorithm 6.3 SEB-MGSI

1: Input: $\{V_k\}, \{\alpha_k\}, \{\beta_k\}, \omega$.
2: **Phase 1:** Initialization
3: Set $Q_k(1)$ as the initial amount of data backlog, $N_k(1) = 0$, $F_k(1) = 0$, $\bar{\theta}_{k,j,0} = 0$, and
   $x_{k,j,t} = 0, j = 1, 2, \cdots, J + 1, \forall m_k \in \mathcal{M}, \forall t \in \mathcal{T}$.
4: **Repeat**
5: **Phase 2:** Preference list construction
6: Each MTD calculates its preference value towards each option as (6.21).
7: Each MTD constructs its preference list $\mathcal{F}_k$ and any $m_k \in \mathcal{M}_t$ transmits $\mathcal{F}_k$ to the edge server
   for iterative matching.
8: **Phase 3:** Iterative matching
9: **Step 1:** Initialization
10: Initialize $\phi = \emptyset, \Omega = \emptyset$.
11: **Step 2:** Pricing-based iterative matching
12: **if** $\exists \phi(m_k) = \emptyset$ **then**
13:     any $m_k \in \mathcal{M}_t$ selects its most preferred subchannel in $\mathcal{F}_k$.
14:     **if** any $c_j \in C$ selected by only one MTD m_k **then**
15:         $\phi(m_k) = c_j$.
16:     **else**
17:         Add c_j into Ω.
18:         **for** $c_j \in \Omega$ **do**
19:             c_j raises its price $\rho_{k,j}$ as (6.22).
20:             All the MTDs selecting c_j update their preferences as (6.21) and renew their selection
                strategies.
21:         **end for**
22:     **end if**
23: **end if**
24: Observe $z_{k,j,t}, E_{k,j,t}$ and whether the delay requirement can be satisfied or not.
25: Update $U_k(t)$ and $Y_k(t + 1)$ based on (6.4) and (6.6).
26: Update $Q_k(t + 1)$, $N_k(t + 1)$, and $F_k(t + 1)$ as (6.1) and (6.15).
27: **Until** $t > T$.

---

initialization (Line 2–3), preference list construction (Line 5–7) and iterative matching (Line 8–22).

(1) *Initialization* The initial length of all the queues and initial values of all the selection indicators are set as zero.

(2) *Preference List Construction* In the second phase of preference list construction, Since the preference of m_k towards any option j, $j = 1, \cdots, J + 1$, is inversely proportional to $\theta_{k,j,t}$, it can be simply expressed as

$$L_{k,j,t} = \frac{1}{\theta_{k,j,t}} - \rho_{k,j}\mathbb{I}\{j < J + 1\}, \tag{6.21}$$

where $\rho_{k,j}$ represents the cost of matching m_k with c_j, the initial value of which is set as zero.

Denote the preference list of m_k towards all the $J + 1$ options as $\mathcal{F}_k$, which is obtained by sorting all the $L_{k,j,t}$, $j = 1, 2, \cdots, J + 1$, in a descending order.

If option $J + 1$ ranks the first in $\mathcal{F}_k$, m_k will skip the iterative matching process and remain idle during this slot. Otherwise, any $m_k \in \mathcal{M}_t$ updates $\mathcal{F}_k$ by removing option $J + 1$, where $\mathcal{M}_t \subseteq \mathcal{M}$ is the set of MTDs selecting to transmit data in the t-th slot. Then any $m_k \in \mathcal{M}_t$ transmits it to the edge server for resolving matching conflicts based on the following procedures:

(3) *Iterative Matching*

Step 1: Initialization

- Initialize $\phi = \emptyset$ and $\Omega = \emptyset$. Here, Ω denotes the conflicting set of subchannels which are selected by more than one MTDs.

Step 2: Pricing-based iterative matching
Repeat

- If $\exists \phi(m_k) = \emptyset$, any $m_k \in \mathcal{M}_t$ selects its most preferred subchannel in $\mathcal{F}_k$.

- For any subchannel $c_j \in C$, if it is selected by only one MTD, e.g., m_k, then they are directly matched, i.e., $\phi(m_k) = c_j$. Otherwise, add c_j into Ω.
- If $\Omega \neq \emptyset$,

  - Each subchannel $c_j \in \Omega$ raises its price $\rho_{k,j}$ as

  $$\rho_{k,j} = \rho_{k,j} + \frac{\Delta \rho_j}{F_k(t)}, \tag{6.22}$$

  where $\Delta \rho_j$ is the step size for price rising.
  - All the MTDs which have selected c_j recalculate their preferences towards c_j based on (6.21), and renew their selection strategies accordingly. If the cost of c_j is too high, some MTDs will give it up and select other subchannels.
  - Repeat the pricing process until only one MTD remains, e.g., m_k. Then, set $\phi(m_k) = c_j$ and remove c_j from Ω.

  - If any c_j in $\mathcal{F}_k$ has been matched with other MTDs and is unavailable to m_k, then $\phi(m_k) = m_k$.

Until $\forall \phi(m_k) \neq \emptyset$.

Finally, the MTDs select the subchannels based on the derived ϕ, observe the corresponding results $z_{k,j,t}$, $E_{k,j,t}$ associated with $x_{k,j,t} = 1$, and whether the delay requirement can be satisfied or not. Then, each MTD m_k updates $U_k(t)$, $Y_k(t + 1)$, $Q_k(t + 1)$, $N_k(t + 1)$, and $F_k(t + 1)$ as (6.19), (6.20), (6.4), (6.6), (6.1) and (6.15). The iterations between the phase of preference list construction and the phase of iterative matching are terminated when $t > T$.

In the pricing-based matching, the price of occupying c_j for m_k is inversely proportional to $F_k(t)$, thereby allowing MTDs with larger service reliability deficit to have a higher probability to be matched with a subchannel, which further enhances service reliability awareness.

6.3.2 SEBC-MUCB Algorithm for the Nonideal Case

In the nonideal case where the nonlocal information required to construct preference lists of MTDs is unavailable, the matching-based SEB-GSI algorithm is infeasible. Following the idea of SEB-UCB developed in Sect. 6.2.3, an intuitive solution is to enable a MTD to estimate its preference list via online learning. We augment SEB-UCB by adding conflict awareness into the learning process, and develop the matching-learning based SEBC-MUCB algorithm. In SEBC-MUCB, a MTD can learn the impacts of decision coupling and matching conflicts by continuously observing the difference between its matching preference and actual matching results.

SEBC-MUCB is summarized in Algorithm 6.4, which consists of three phases, i.e., initialization (Line 2–4), pricing-based matching (Line 6–9), and learning (Line 10–14).

In the first phase of initialization, firstly, the initial length of all the queues and initial values of all the selection indicators are set as zero. Then, for any $m_k \in \mathcal{M}$, it is temporarily matched with every $c_j \in \mathcal{C}$ to observe the performances of throughput, energy consumption and delay.

In the second phase of pricing-based matching, m_k estimates its preference towards the j-th option as

$$\widetilde{L}_{k,j,t} = \frac{1}{\bar{\theta}_{k,j,t-1}} + \omega \sqrt{\frac{2 \ln t}{\hat{x}_{k,j,t-1}}} - \rho_{k,j}\mathbb{I}\{j < J+1\}. \qquad (6.23)$$

Here, the preference value of m_k towards an option that has not been selected, e.g., $\hat{x}_{k,j,t-1} = 0$, is defined as $+\infty$ so that each option can be selected by m_k at least once.

Then, based on (6.23), m_k constructs its preference list $\mathcal{F}_k$ similarly as Sect. 6.3.1 and transmits it to the edge server. Next, MTDs are matched with subchannels based on the pricing-based matching. Eventually, each MTD selects the subchannel according to the obtained $\phi(m_k)$.

In the third phase of learning, each MTD m_k observes the corresponding results $z_{k,j,t}$, $E_{k,j,t}$ associated with $x_{k,j,t} = 1$ and whether the delay requirement can be satisfied or not. Then, each MTD m_k updates $\bar{\theta}_{k,j,t}$, $\hat{x}_{k,j,t}$, $U_k(t)$, $Y_k(t+1)$, $Q_k(t+1)$, $N_k(t+1)$, and $F_k(t+1)$ as (6.19), (6.20), (6.4), (6.6), (6.1) and (6.15). The iterations between the phase of pricing-based matching and the phase of learning are terminated when $t > T$.

6.4 Performance Results and Discussions

In this section, we validate the algorithms via simulations under the scenarios of single-MTD and multi-MTD, respectively.

Algorithm 6.4 SEBC-MUCB

1: Input: $\{V_k\}$, $\{\alpha_k\}$, $\{\beta_k\}$, ω.
2: **Phase 1:** Initialization
3: Set $Q_k(1)$ as the initial amount of data backlog, $N_k(1) = 0$, $F_k(1) = 0$, $\bar{\theta}_{k,j,0} = 0$, $\hat{x}_{k,j,0} = 0$, and $x_{k,j,t} = 0$, $j = 1, 2, \cdots, J + 1$, $\forall m_k \in M$, $\forall t \in \mathcal{T}$.
4: Temporarily match $\forall m_k \in M$ with $\forall c_j \in C$ to observe the performances of throughput, energy consumption and delay.
5: **Repeat**
6: **Phase 2:** Pricing-based matching
7: Each MTD calculates its preference value towards each option as (6.23).
8: Each MTD constructs its preference list $\mathcal{F}_k$ and the $m_k \in M_k$ transmits $\mathcal{F}_k$ to the edge server for iterative matching.
9: Each MTD performs the corresponding selection based on $\phi(m_k)$.
10: **Phase 3:** Learning
11: Observe $z_{k,j,t}$, $E_{k,j,t}$ and whether the delay requirement can be satisfied or not.
12: Update $\bar{\theta}_{k,j,t}$ and $\hat{x}_{k,j,t}$ based on (6.19) and (6.20).
13: Update $U_k(t)$ and $Y_k(t + 1)$ based on (6.4) and (6.6).
14: Update $Q_k(t + 1)$, $N_k(t + 1)$, and $F_k(t + 1)$ as (6.1), (6.15).
15: **Until** $t > T$.

6.4.1 Performance Under the Single-MTD Scenario

In the single-MTD scenario, we consider one MTD and three subchannels over a total period of $T = 10^3$ time slots, i.e., $K = 1$ and $J = 3$. We set $\tau = 1$ s, $P_{TX} = 1$ W, $E_{k,max} = 700$ J. We assume that $A_k(t)$ follows a uniform distribution within the interval $[0.9\bar{A}_k, 1.1\bar{A}_k]$ Mbits, where $\bar{A}_k = 20$ Mbits represents the time-average amount of collected data. The initial value $Q_k(1)$ is randomly selected within the interval $[0.8\bar{A}_k, 1.2\bar{A}_k]$ Mbits. The computational complexity is set as $\lambda_{k,t} = 10^3$ CPU cycles/bit. The available computational resource for m_k in the t-th slot $\xi_{k,t}$ is randomly distributed within the interval $[0.9\bar{\xi}_k, 1.1\bar{\xi}_k]$ CPU cycles, where $\bar{\xi}_k = 18 \times 10^9$ CPU cycles represents the time-average amount of computational resource. $U_k(t)$ does not need to be initialized, the value of which depends on the selection strategies, CSI as well as local data backlog. The service reliability requirement is set as $\eta_k = 0.7$. We set $V_k = 1$, $\alpha_k = 5$, and $\beta_k = 3$ to balance the tradeoff among throughput performance, energy consumption, and service reliability. The achievable transmission rate of subchannel s_j in each slot follows a uniform distribution within the range $[0.8\bar{R}_j, 1.2\bar{R}_j]$, where $\bar{R}_j$ represents the average transmission rate. We set $\bar{R}_j = 10, 20, 30$ Mbits when $j = 1, 2, 3$.

The weight of exploration ω is set as 1. Two heuristic algorithms are used for comparison. One is the conventional UCB algorithm introduced in [85], and the other is the random selection algorithm in which m_k randomly selects a subchannel at each slot. The SEB-GSI with perfect GSI is used as an upper performance benchmark.

Figure 6.2a and b show the cumulative network throughput and cumulative energy consumption performances over a total of 10^3 slots. Compared with UCB and random selection, the SEB-UCB with only local information can improve

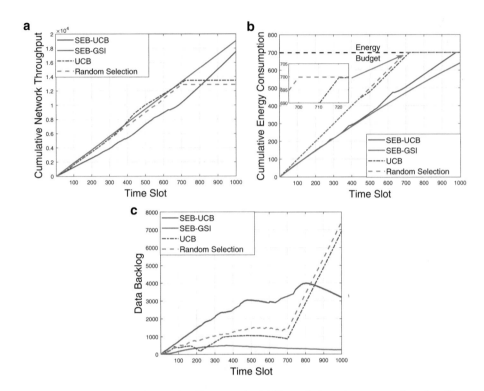

Fig. 6.2 Performances under single-MTD scenario. (**a**) Network throughput versus time slot. (**b**) Energy consumption versus time slot. (**c**) Data backlog versus time slot

throughput by 30 and 36% respectively, while satisfying the constraint of energy consumption. Particularly, there exists a performance floor after 700 slots. The reason is demonstrated in Fig. 6.2b, which explicitly shows that the two heuristic algorithms use energy more aggressively at the beginning and then run out of the energy at $t = 700$ and $t = 720$, thereby leaving no energy for data transmission. It is noted that the energy consumption of the algorithms will not increase after $t = 1000$ since the energy budget is exactly exhausted at $t = 1000$, i.e., the algorithms can well exploit the available energy during the specified optimization duration compared with other heuristic algorithms. Besides, the SEB-UCB performs just slightly worse than the SEB-GSI algorithm with perfect GSI. The curve trends of both the network throughput and energy consumption performances track those of SEB-GSI strictly.

Figure 6.2c shows the data backlog performance. Simulation results demonstrate that SEB-UCB can provide bounded data backlog, while the backlogs of UCB and random selection increase linearly with time after 700 slots, which significantly degrades the queue stability performance and may even lead to severe data loss.

6.4.2 Performance Under the Multi-MTD Scenario

For the multi-MTD scenario, we consider three MTDs and three subchannels, i.e., $K = J = 3$. We set $E_{k,max} = 730\,\text{J}$, $\eta_k = 0.73$, $V_k = 1$, $\alpha_k = 20$, and $\beta_k = 25$, $\forall m_k \in \mathcal{M}$. The other simulation parameters remain the same as those in the single-MTD scenario.

Five heuristic algorithms are used for comparison. The first one is the EBC-MUCB algorithm without service reliability awareness, i.e., the service reliability constraint is not considered. The second one is the SBC-MUCB algorithm without energy awareness, i.e., the energy consumption constraint is not considered. The third one is the conventional UCB algorithm, and the fourth one is random selection. The fifth one is the Lyapunov optimization-based access control and resource allocation (ACRA) algorithm developed in [54]. ACRA requires perfect GSI to find the optimum option. Here, we assume that only the CSI of the previous slot is available, i.e., the CSI is outdated information. In other words, optimization at the t-th slot is performed based on the CSI of the $(t - 1)$-th slot.

Figure 6.3a shows the cumulative network throughput versus time slot. The SEBC-MUCB outperforms UCB and random selection by 13.7% and 31.2%,

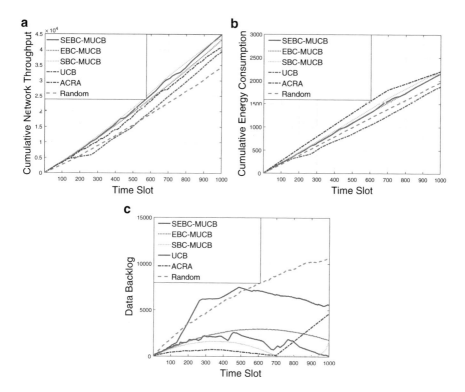

Fig. 6.3 Performances under multi-MTD scenario. (**a**) Network throughput versus time slot. (**b**) Energy consumption versus time slot. (**c**) Data backlog versus time slot

respectively. Compared with SBC-MUCB and EBC-MUCB, SEBC-MUCB improves throughput by 3.46% and 3.96%, respectively, due to the additional consideration of energy awareness and service reliability awareness. Taking SBC-MUCB as an example, although it achieves a higher throughout at the beginning, it runs out of energy at $t = 981$ and is forced to be idle for the remaining slots, which significantly degrades the overall throughput performance.

Figure 6.3b shows the cumulative energy consumption versus time slot. Simulation results show that The energy consumption of SEBC-MUCB and EBC-MUCB algorithms has not exceeded the energy budget due to energy awareness. Different from the scenario of single-MTD, UCB consumes the least energy since the frequent selection conflicts force MTDs to remain idle so that the energy consumption becomes less.

Figure 6.3c demonstrates that SEBC-MUCB achieves the least data backlog among all the algorithms. In comparison, the data backlog of SBC-MUCB increases dramatically after $t = 981$ due to the ignorance of energy awareness. UCB performs worse than EBC-MUCB and SBC-MUCB since the frequent selection conflicts impede MTDs from data transmission and data backlog becomes very large.

Figure 6.4 shows the service reliability deficit versus time slot. The SEBC-MUCB can meet the service reliability requirement and achieve the second least service reliability deficit. Although SBC-MUCB achieves the least service reliability deficit, its throughput and energy consumption performance are worse than SEBC-MUCB because only service reliability awareness is considered. The service reliability deficit of EBC-MUCB increases dramatically after $t = 700$ due to the

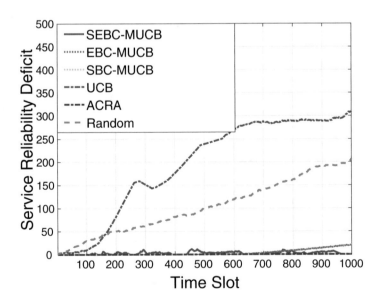

Fig. 6.4 Service reliability deficit versus time slot

negligence of service reliability awareness. UCB performs the worst since it has not been endowed with the capability of conflict resolution.

From Figs. 6.3a to 6.4, we can find that although the energy consumption and the service reliability deficit of ACRA are nearly the same as those of SEBC-MUCB, the throughput performance and the data backlog performance are worse. SEBC-MUCB outperforms ACRA by 10.58% in terms of throughput, and 4783.76% in terms of data backlog due to the endowed capability of online learning. Particularly, the data backlog performance of ACRA is significantly degraded by employing the outdated CSI for optimization. Therefore, we can conclude that learning plays an important role for backlog reduction under the scenario where perfect GSI is unavailable.

Figure 6.5 shows the impact of parameter V_k on the throughput performances of MTDs. Specifically, we set $V_1 = V_2 = 1$ for m_1 and m_2, while V_3 increases from 10^{-4} to 10^3 for m_3. Simulation results demonstrate that as V_3 increases, the throughput of m_3 increases first and then decreases, while the throughput of m_1 shows the opposite trend. The rationale is that when V_3 increases from 10^{-4} to 25 ($\log(V_3)$ increase from -4 to 1.4), m_3 puts a larger weight on the throughput, and becomes more active to explore channels for throughput improvement. This will cause more channel selection conflicts, thereby reducing the throughput of m_1. However, when V_3 is too large ($\log(V_3) > 1.4$), m_3 over-evaluates throughput and has little concern on energy consumption. It will quickly run out of energy and is forced to remain idle, which significantly degrades the throughput performance.

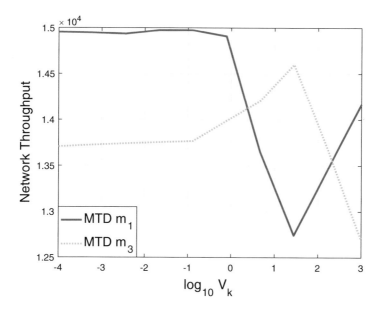

Fig. 6.5 Impact of V_k

Meanwhile, other MTDs such as m_1 can benefit from the idle state of m_3 since the channel selection conflicts is relieved.

In this chapter, we proposed learning-based channel selection which incorporates service reliability awareness, energy awareness and backlog awareness. We started from single-MTD scenario and demonstrated distributed low-complexity SEB-GSI algorithm with CSI and SEB-UCB algorithm under information uncertainty. Then, we extended it to the multi-MTD scenario and introduced SEBC-MUCB algorithm by integrating MAB, Lyapunov optimization and matching theory. Simulation results demonstrate that the SEB-UCB can improve throughput by 30 and 36% compared with UCB and random selection. SEBC-MUCB outperforms UCB and random selection by 13.7 and 31.2% while stabilizing data backlog queue and satisfying energy consumption constraint as well as service reliability requirement. Due to the limited computational capability and battery capacity of MTDs, we only consider the scenario of task offloading, while local computing is ignored. The future work will focus on the online cross-layer resource optimization including local computation, rate control, channel selection, and resource allocation in the edge server under information uncertainty.

Chapter 7
Licensed and Unlicensed Spectrum Management for Energy-Efficient Cognitive M2M

7.1 Framework of CM2M Network

In this section, the system model and the problem formulation are introduced.

7.1.1 System Model

Figure 7.1 shows a single-cell CM2M network which is composed of a LTE-U base station (LBS), an edge server, and a massive number of MTDs. The edge server is co-located with the LBS to provide computing services for MTDs within the cell, while the LBS provides communication services.

We consider the scenario where the licensed spectrum coexists with the unlicensed spectrum. There are a total of J orthogonal subchannels including both the licensed spectrum used for LTE and the unlicensed spectrum used for LTE-U, the set of which is defined as $C = \{c_1, \cdots, c_j, \cdots, c_J\}$. The bandwidth of subchannel c_j is denoted as B_j. Specifically, the set of licensed subchannels is defined as $C_L = \{c_1, \cdots, c_l, \cdots, c_L\}$, which corresponds to the first L elements of set C. The set of unlicensed subchannels is defined as $C_U = \{c_{L+1}, \cdots, c_{L+u}, \cdots, c_{L+U}\}$, which corresponds to the last U elements of set C, i.e., $L + U = J$.

According to the delay requirements, MTDs can be divided into two categories. Delay-sensitive MTDs such as leakage detectors are classified as primary users (PUs), and the set of N PUs is denoted as $\mathcal{PU} = \{PU_1, \cdots, PU_n, \cdots, PU_N\}$. On the other hand, delay-tolerant MTDs such as smart meters are classified as secondary users (SUs) [86], and the set of K SUs is denoted as $\mathcal{SU} = \{SU_1, \cdots, SU_k, \cdots, SU_K\}$ ($K \gg J$). PUs are allocated with the licensed spectrum to guarantee reliable service provisioning. In comparison, SUs transmit data by either opportunistically leveraging the licensed subchannels that are temporarily unoccupied by PUs, or exploiting the unlicensed subchannels. We assume that

© Springer Nature Switzerland AG 2021
Z. Zhou et al., *Green Internet of Things (IoT): Energy Efficiency Perspective*,
Wireless Networks, https://doi.org/10.1007/978-3-030-64054-5_7

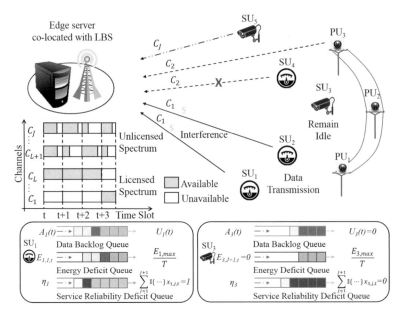

Fig. 7.1 The CM2M network

an SU is required to pay some fees for utilizing a licensed subchannel, while the unlicensed subchannels are free. Denoting the price of subchannel c_j as r_j (cent/Mbit), we have $r_j = 0$ if $j = L+1, L+2, \cdots, J$. Multiple SUs with the same priority can access idle subchannels simultaneously. In this case, an SU will receive interference from other SUs selecting the same subchannel.

A time-slotted model is adopted where the total time period is divided into T slots, the set of which is defined as $\mathcal{T} = \{1, \cdots, t, \cdots, T\}$ [87]. The channel CSI and channel availability are assumed to be constant within a slot but vary across different slots. In each slot, an SU determines its channel selection strategy individually. The set of channel selection indicators consists of $J+1$ binary elements, which is denoted as $\{x_{k,j,t}\}$, where $x_{k,j,t} \in \{0, 1\}$. When $j = 1, 2, \cdots, J$, $x_{k,j,t} = 1$ represents that SU_k selects c_j for data transmission in the t-th slot and when $j = J+1$, $x_{k,j,t} = 1$ represents that SU_k remains idle. For licensed subchannels, denote channel availability as L binary elements $\{a_{j,t}\}$. When $j = 1, 2, \cdots, L$, $a_{j,t} = 1$ represents that c_j is available for SUs during the t-th slot and $a_{j,t} = 0$ otherwise. Due to the random behaviors of PUs, the prior knowledge of $\{a_{j,t}\}$ is unknown to SUs. For an unlicensed subchannel c_j, $j = L+1, L+2, \cdots, J$, its availability is specified by duty cycle $\varpi_{j,t}$ [88], i.e., the ratio of channel availability duration to the entire slot, which was originally designed to enable the coexistence between LTE and Wi-Fi. It is assumed that the information of duty cycle is broadcast to SUs by the LBS at the beginning of each slot. Therefore, the duty cycle $\varpi_{j,t}$ is possessed by the SUs.

An example is shown in Fig. 7.1, where SU_1 and SU_2 select c_1 and the simultaneous selection causes interference to each other. Since c_1 is a licensed subchannel, both SU_1 and SU_2 should pay for utilizing c_1. SU_3 remains idle. SU_4 cannot use c_2 because it is currently occupied by PU_3. SU_5 selects an unlicensed subchannel, which is free.

In the following, the models of task transmission, energy consumption, delay, and service reliability are introduced.

(1) Task Transmission Model

In the t-th slot, $A_k(t)$ new tasks with equal size φ_k arrive at $SU_k \in \mathcal{SU}$, which are firstly stored in the local buffer and then are transmitted to the edge server. The statistical information of $A_k(t)$ is unknown. The total task size is $\varphi_k A_k(t)$. Meanwhile, SU_k has to retransmit $Y_k(t+1)$ amount of data which have not been correctly delivered due to bit error. The task data stored in the buffer of SU_k can be modeled as a queue, i.e., queue k, the backlog of which is dynamically evolved as

$$Q_k(t+1) = \max\{Q_k(t) + \varphi_k A_k(t) - U_k(t), 0\} + Y_k(t+1), \tag{7.1}$$

where $U_k(t)$ denotes the amount of data transmitted to the edge server by SU_k in the t-th slot.

Uplink transmission is considered here. Denote $g_{k,j,t}$ as the uplink channel gain of c_j between the SU_k and the LBS. Given $x_{k,j,t}$, the signal to interference plus noise ratio (SINR) and the achievable transmission rate of SU_k are given by

$$\gamma_{k,j,t} = \begin{cases} \dfrac{P_{TX}g_{k,j,t}}{\sum_{k'\neq k} x_{k',j,t} P_{TX}g_{k',j,t}+\delta^2}, & j = 1, 2, \cdots, J, \\ 0, & j = J+1, \end{cases}$$

$$R_{k,j,t} = B_j \log_2(1 + \gamma_{k,j,t}), \tag{7.2}$$

where δ^2 is the noise power, and P_{TX} is the transmission power. The throughput of SU_k choosing option j in the t-th slot is given by

$$z_{k,j,t} = \begin{cases} a_{j,t} \min\{Q_k(t), \tau R_{k,j,t}\}, & j = 1, 2, \cdots, L. \\ \min\{Q_k(t), \varpi_{j,t}\tau R_{k,j,t}\}, & j = L+1, L+2, \cdots, J. \\ 0, & j = J+1. \end{cases} \tag{7.3}$$

For SU_k, the cost to utilize c_j is defined as the unit price r_j multiplied by the throughput $z_{k,j,t}$, i.e., $r_j z_{k,j,t}$. The amount of data transmitted to the edge server is

$$U_k(t) = \sum_{j=1}^{J+1} x_{k,j,t} z_{k,j,t}. \tag{7.4}$$

Denote the BER for SU_k transmitting data through c_j in the t-th slot as $P^e_{k,j,t}$. The coherent BPSK modulation is adopted. The BER [75] and the amount of data retransmitted in the next slot is

$$P^e_{k,j,t} = \frac{1}{2}\text{erfc}\sqrt{\gamma_{k,j,t}},$$

$$Y_k(t+1) = U_k(t)P^e_{k,j,t}. \tag{7.5}$$

(2) Energy Consumption Model

In the t-th slot, the energy consumption of SU_k is derived as

$$E_{k,j,t} = \begin{cases} a_{j,t}P_{\text{TX}}\min\{\frac{Q_k(t)}{R_{k,j,t}}, \tau\}, & j = 1, 2, \cdots, L. \\ P_{\text{TX}}\min\{\frac{Q_k(t)}{R_{k,j,t}}, \varpi_{j,t}\tau\}, & j = L+1, L+2, \cdots, J. \\ 0, & j = J+1. \end{cases} \tag{7.6}$$

Therefore, the long-term energy consumption of SU_k must satisfy

$$E_{k,T} = \sum_{t=1}^{T}\sum_{j=1}^{J+1} x_{k,j,t}E_{k,j,t} \leq E_{k,max}, \tag{7.7}$$

where $E_{k,max}$ denotes the energy budget of SU_k over T slots.

(3) Delay Model

The total offloading delay is the sum of the transmission delay and the computational delay, which is given by $d^{total}_{k,j,t} = d^{tra}_{k,j,t} + d^{com}_{k,j,t}$. Given $x_{k,j,t}$, the transmission delay is given by

$$d^{tra}_{k,j,t} = \begin{cases} a_{j,t}\min\{\frac{Q_k(t)}{R_{k,j,t}}, \tau\}, & j = 1, 2, \cdots, L. \\ \min\{\frac{Q_k(t)}{R_{k,j,t}}, \varpi_{j,t}\tau\}, & j = L+1, L+2, \cdots, J. \\ +\infty, & j = J+1. \end{cases} \tag{7.8}$$

Denoting the computational intensity of the task data transmitted by SU_k in the t-th slot as $\lambda_{k,t}$ (CPU cycles/Mbit), it requires $z_{k,j,t}\lambda_{k,t}$ CPU cycles to process the task data. Denoting the available computational resources for SU_k in the t-th slot as $\xi_{k,t}$ (CPU cycles/s), the computational delay is calculated as

$$d^{com}_{k,j,t} = \begin{cases} \frac{z_{k,j,t}\lambda_{k,t}}{\xi_{k,t}}, & j = 1, 2, \cdots, J, \text{ and } a_{j,t} = 1. \\ +\infty, & j = J+1, \text{ or } a_{j,t} = 0. \end{cases} \tag{7.9}$$

(4) Service Reliability Requirement Model

The service reliability requirement is modeled in terms of delay, which is defined as $d_{k,t}$. The task offloading is considered to be a failure if the offloaded task cannot be processed within the specified delay requirement, i.e., $d_{k,j,t}^{total} > d_{k,t}$. Denote $X_{k,T}$ as the number of successful task offloading for SU_k over T slots, which is

$$X_{k,T} = \sum_{t=1}^{T} \sum_{j=1}^{J+1} \mathbb{I}\{d_{k,j,t}^{total} \le d_{k,t}\} x_{k,j,t}. \tag{7.10}$$

$\mathbb{I}\{x\}$ is an indicator function with $\mathbb{I}\{x\} = 1$ if event x is true and $\mathbb{I}\{x\} = 0$ otherwise. The results of $\mathbb{I}\{d_{k,j,t}^{total} \le d_{k,t}\}$ are determined by the self-optimizing CPU allocation of the edge server and are fed back to SU_k at the end of the t-th slot.

The service reliability requirement is defined as

$$\frac{X_{k,T}}{T} \ge \eta_k, \tag{7.11}$$

where $\eta_k \in (0, 1]$ represents the minimum successful probability of task offloading.

7.1.2 Problem Formulation

The objective is to maximize the total utility of SUs in the network subject to the long-term constraints of energy consumption and service reliability. The utility of SU_k can be defined as

$$W_{k,j,t} = \begin{cases} z_{k,j,t} - r_j z_{k,j,t}, & j = 1, 2, \cdots, J. \\ 0, & j = J + 1. \end{cases} \tag{7.12}$$

The corresponding utility maximization problem for SUs is formulated as

$$\max_{\{x_{k,j,t}\}} \sum_{k=1}^{K} \sum_{t=1}^{T} \sum_{j=1}^{J+1} x_{k,j,t} W_{k,j,t},$$

$$\text{s.t. } C_1 : \sum_{j=1}^{J+1} x_{k,j,t} = 1, \forall SU_k \in \mathcal{SU}, \forall t \in \mathcal{T},$$

$$C_2 : \sum_{t=1}^{T} \sum_{j=1}^{J+1} x_{k,j,t} E_{k,j,t} \le E_{k,max}, \forall SU_k \in \mathcal{SU},$$

$$C_3 : \frac{X_{k,T}}{T} \ge \eta_k, \forall SU_k \in \mathcal{SU}, \tag{7.13}$$

where C_1 is the channel selection constraint, i.e., an SU can select only one subchannel at most or remains idle in each slot. C_2 and C_3 correspond to the constraints of energy consumption and service reliability, respectively.

7.2   Context-Aware Learning-Based Channel Selection for CM2M

In this section, the context-aware learning-based channel selection algorithm is introduced. We start from the ideal scenario with GSI, and develop a context-aware channel selection algorithm with GSI named C^2-GSI. Then, we extend C^2-GSI to the nonideal scenario with only local information, and present the C^2-EXP3 algorithm.

7.2.1   Problem Transformation

To solve the problem defined in (7.13), the long-term constraints C_2 and C_3 can be transformed to queue stability constraints based on the concept of virtual queue [87]. We define the virtual queue of energy deficit as $N_k(t)$ and the virtual queue of service reliability deficit as $F_k(t)$, which are evolved as

$$N_k(t+1) = \max\{N_k(t) + \sum_{j=1}^{J+1} x_{k,j,t} E_{k,j,t} - \frac{E_{k,max}}{T}, 0\},$$

$$F_k(t+1) = \max\{F_k(t) + \eta_k - \sum_{j=1}^{J+1} \mathbb{I}\{d_{k,j,t}^{total} \le d_{k,t}\} x_{k,j,t}, 0\}, \qquad (7.14)$$

where $N_k(0) = F_k(0) = 0$. $N_k(t)$ and $F_k(t)$ represent the energy consumption deviation and the service reliability deviation, respectively. Figure 7.1 explicitly shows the examples of queue evolution. For each SU, e.g., SU_k, there exist three queues, i.e., the data backlog queue $Q_k(t)$, energy deficit queue $N_k(t)$, and service reliability deficit queue $F_k(t)$. The queues are dynamically updated in each slot based on (1) and (14). Comparing SU_1 with SU_3, due to the larger $Q_1(t)$ and $F_1(t)$, SU_1 selects the licensed subchannel c_1 for task offloading. Due to the larger $N_3(t)$, SU_3 remains idle to reduce energy consumption.

Based on Lyapunov optimization [83], if the energy consumption of SU_k does not exceed the energy budget, an online optimization problem is defined at each slot to maximize the utility and service reliability while minimizing energy consumption, which is given by

$$\max_{\{x_{k,j,t}\}} \sum_{k=1}^{K} \sum_{j=1}^{J+1} x_{k,j,t} \theta_{k,j,t},$$

$$\text{s.t.} \ C_1, \tag{7.15}$$

where

$$\theta_{k,j,t} = V_k W_{k,j,t} - \alpha_k N_k(t) E_{k,j,t}$$

$$+ \beta_k F_k(t) \Big(\sum_{j=1}^{J+1} \mathbb{I}\{d_{k,j,t}^{total} \leq d_{k,t}\} x_{k,j,t} - \eta_k \Big). \tag{7.16}$$

Here, $\theta_{k,j,t}$ is a weighted sum of the utility, energy consumption and service reliability, where V_k, $\alpha_k N_k(t)$, and $\beta_k F_k(t)$ are the corresponding positive weights.

It is noted that problems defined in (7.15) and (7.13) may not be equal. Nevertheless, we will prove later that the results of the problem defined in (7.15) are within a bounded deviation from the optimal results. In addition, C_2 can be guaranteed by defining that if the energy budget of SU_k is exhausted, then it is forced to remain idle and is not able to transmit data. In other words, in the t-th slot, the problem defined in (7.15) will be solved if and only if the energy budget is not exhausted.

We adopt the information classification method introduced in [89]. The information required to solve the problem defined in (7.15) can be classified into two categories depending on whether the SUs process the information or not:

- **Local Information:** information that is possessed by SU_k, e.g., the queue backlog $Q_k(t)$, transmission power P_{TX}, total energy budget $E_{k,max}$, computational intensity of task data $\lambda_{k,t}$, task delay requirement $d_{k,t}$, duty cycle $\varpi_{j,t}$, minimum successful probability of task offloading η_k, and unit price r_j of c_j.
- **Nonlocal Information:** information that is not possessed by SU_k, e.g., the uplink channel gain $g_{k,j,t}$, available computational resources of the edge server $\xi_{k,t}$, channel availability $a_{j,t}$, and channel selection strategies of other SUs $x_{i,j,t}$, $SU_i \in \mathcal{SU}, i \neq k$.

Based on whether SU_k has the nonlocal information or not, we consider both ideal and nonideal scenarios. In the ideal scenario, SU_k has the perfect knowledge of GSI, which includes both local and nonlocal information. In the nonideal scenario, SU_k only knows the local information while the nonlocal information is unavailable.

7.2.2 C^2-GSI for Channel Selection with GSI

For the ideal scenario with GSI, we introduce a context-aware channel selection algorithm named C^2-GSI with service reliability awareness, energy awareness and

---

Algorithm 7.1 C^2-GSI

1: Input: $\{V_k\}, \{\alpha_k\}, \{\beta_k\}$.
2: **Phase 1:** Initialization
3: Set $Q_k(1)$ as the initial amount of data backlog, $\{x_{k,j,t}\} = 0$, $N_k(0) = F_k(0) = 0$, $\mathrm{SU}_k \in \mathcal{SU}$.
4: **Repeat**
5: **Phase 2:** Decision making
6: Input: $\{g_{k,j,t}\}, \delta^2$.
7: Calculate the accurate value of $\{\theta_{k,j,t}\}$ with GSI.
8: Choose $\phi_{k,t}$ by solving the problem defined in (7.15).
9: Update $Q_k(t+1)$, $N_k(t+1)$, and $F_k(t+1)$ as (7.1) and (7.14).
10: **Until** $t > T$.

---

backlog awareness. C^2-GSI does not require future non-causal information. The detailed procedures are summarized in Algorithm 7.1, which consists of two phases, i.e., initialization (Line $2 \sim 3$) and decision making (Line $5 \sim 9$).

In the initialization phase, set $Q_k(1)$ as the initial amount of data backlog. The initial length of all the virtual queues and all the selection indicators are set as zero.

Then, the decision making phase is executed in a slot-by-slot fashion. At the beginning of the t-th slot, m_k calculates the value of $\theta_{k,j,t}$ towards option j, $j = 1, 2, \cdots, J + 1$, based on the current GSI. Denote the decision of SU_k in the t-th slot as $\phi_{k,t}$. $\phi_{k,t}$ can be found by solving the problem defined in (7.15), which is equivalent to a minimum seeking problem with computational complexity $O(J)$. Afterwards, SU_k sets $x_{k,\phi_{k,t},t} = 1$, and updates all queues accordingly. In the next slot, the iteration continues until $t > T$.

Context awareness is achieved through the dynamic adjustment of channel selection strategy based on the values of $F_k(t)$, $N_k(t)$, and $Q_k(t)$. Details are given as follows:

- **Service reliability awareness:** When the service reliability deviates severely from the service reliability requirement, a large weight $F_k(t)$ will be placed on the service reliability term which enforces SU_k to select the option with higher successful chances of task offloading, thereby enabling service reliability awareness.
- **Energy awareness:** When the energy consumption significantly exceeds the current energy budget, a large weight $N_k(t)$ on the energy consumption term will enforce SU_k to select the option with less energy consumption, e.g., remaining idle, thereby enabling energy awareness.
- **Backlog awareness:** A large data backlog $Q_k(t)$ will lead to a large throughput, e.g., for licensed spectrum, $z_{k,j,t} = \tau R_{k,j,t}$ based on (7.3), which motivates SU_k to choose the subchannel with higher data transmission rate, thereby enabling backlog awareness.

7.2.3 C^2-EXP3 for Channel Selection with Local Information

C^2-GSI cannot be applied in the nonideal scenario because the nonlocal information are unavailable. EXP3 is a powerful tool to address the adversarial MAB problem with local information and is proved to be Hannan-consistent [90], i.e., as time elapses, the performance difference between EXP3 and always selecting the best single option in hindsight converges to zero with probability 1. Instead of directly calculating $W_{k,j,t}$, EXP3 chooses an option according to a Gibbs distribution, which is a mixture of a uniform distribution ρ and an empirical performance-related distribution $\omega_{k,j,t}$. The uniform distribution ensures that EXP3 keeps exploring occasionally and discovers the potential better options which were non-optimal previously. ρ is initialized within the closed interval [0, 1], which reflects the preference towards exploration. When $\rho \neq 1$, a performance-related weight is used to ensure that the option with the currently optimal performance will be selected with the largest probability, i.e., exploitation. The detailed procedures are summarized in Algorithm 7.2, which consists of three phases, i.e., initialization (Line 2 $\sim$ 3), decision making (Line 5 $\sim$ 7), and learning (Line 8 $\sim$ 10).

In the first phase of initialization, the performance-related weights are initialized as 1, i.e., $\{\omega_{k,j,t}\} = 1$. All the selection indicators and utilities are initialized as 0, i.e., $\{x_{k,j,t}\} = 0$ and $\{W_{k,j,t}\} = 0$.

In the second phase of decision making, for any $S\mathcal{U}_k \in \mathcal{SU}$, the probability of selecting option j is estimated as

$$p_{k,j,t} = (1 - \rho)\frac{\omega_{k,j,t}}{\sum_{j=1}^{J+1} \omega_{k,j,t}} + \frac{\rho}{J + 1}. \tag{7.17}$$

$\phi_{k,t}$ is selected randomly according to the distribution of $p_{k,j,t}$.

In the third phase of learning, SU_k observes $W_{k,\phi_{k,t},t}$ and the task offloading result $\mathbb{I}\{d_{k,\phi_{k,t},t}^{total} \leq d_{k,t}\}$. To compensate the reward of options with infrequent selection, EXP3 defines the estimated reward as the actual reward $W_{k,j,t}$ divided by the probability of selecting option j, which is given by

$$\hat{W}_{k,j,t} = \frac{W_{k,j,t}}{p_{k,j,t}}. \tag{7.18}$$

The estimated reward guarantees that the expected reward of the selected option is equal to the actual reward. Next, SU_k updates $\omega_{k,j,t+1}$ as

$$\omega_{k,j,t+1} = \omega_{k,j,t} \exp(\frac{\rho \hat{W}_{k,j,t}}{J + 1}). \tag{7.19}$$

Since the values of $\{W_{k,j,t}\}$ corresponding to the J unselected options are set to 0, the estimated rewards $\{\hat{W}_{k,j,t}\}$ of the unselected options are 0 based on (7.18). Therefore, the values of $\{\omega_{k,j,t+1}\}$ corresponding to the J unselected options remain

---

Algorithm 7.2 EXP3

1: Input: $\rho \in [0, 1]$.
2: **Phase 1:** Initialization
3: Initialize $\{\omega_{k,j,t}\} = 1$, $\{W_{k,j,t}\} = 0$, $\{x_{k,j,t}\} = 0$.
4: **Repeat**
5: **Phase 2:** Decision making
6: Estimate selection probability of each option as (7.17).
7: Draw channel selection decision $\phi_{k,t}$.
8: **Phase 3:** Learning
9: Observe $W_{k,\phi_{k,t},t}$ and task offloading result.
10: Update $\hat{W}_{k,j,t}$ and $\omega_{k,j,t}$ based on (7.18) and (7.19).
11: **Until** $t > T$.

Algorithm 7.3 C^2-EXP3

1: Input: $\{V_k\}$, $\{\alpha_k\}$, $\{\beta_k\}$, $\rho \in [0, 1]$.
2: **Phase 1:** Initialization
3: Set $Q_k(1)$ as the initial amount of data backlog, $N_k(0) = F_k(0) = 0$, SU$_k \in \mathcal{SU}$. $\{\omega_{k,j,t}\} = 1$, $\{\theta_{k,j,t}\} = 0$, $\{x_{k,j,t}\} = 0$,
4: **Repeat**
5: **Phase 2:** Decision making
6: Estimate selection probability of each option as (7.17).
7: Draw channel selection decision $\phi_{k,t}$.
8: **Phase 3:** Learning
9: Observe $W_{k,\phi_{k,t},t}$ and task offloading result. Calculate $\theta_{k,\phi_{k,t},t}$ as a reward based on (7.16).
10: Update $\hat{\theta}_{k,j,t}$ and $\omega_{k,j,t}$ based on (7.20) and (7.21).
11: Update $Q_k(t + 1)$, $N_k(t + 1)$, and $F_k(t + 1)$ as (7.1) and (7.14).
12: **Until** $t > T$.

---

unchanged based on (7.19). The iteration between decision making and learning continues until $t > T$.

However, with limited energy budget and stringent service reliability requirement, EXP3 will suffer from significant performance degradation due to the context unawareness.

Therefore, we augment EXP3 with context awareness, and introduce C^2-EXP3. The detailed procedures are summarized in Algorithm 7.3, which consists of three phases, i.e., initialization (Line 2 $\sim$ 3), decision making (Line 5 $\sim$ 7), and learning (Line 8 $\sim$ 11).

In the first phase of initialization, set $Q_k(1)$ as the initial amount of data backlog. The performance-related weights are initialized as 1, i.e., $\{\omega_{k,j,t}\} = 1$. All the virtual queue backlogs, selection indicators, and weighted sums of the utility, energy consumption, and service reliability are initialized as 0.

The second phase of C^2-EXP3 is exactly the same as that of EXP3.

In the third phase of learning, SU$_k$ observes $W_{k,\phi_{k,t},t}$ as well as the task offloading result and calculates the reward $\theta_{k,\phi_{k,t},t}$. C^2-EXP3 calculates the estimated reward as

$$\hat{\theta}_{k,j,t} = \frac{\theta_{k,j,t}}{p_{k,j,t}}. \tag{7.20}$$

Next, SU_k updates $\omega_{k,j,t+1}$ as

$$\omega_{k,j,t+1} = \omega_{k,j,t} \exp(\frac{\rho \hat{\theta}_{k,j,t}}{J+1}). \tag{7.21}$$

Since the values of $\{\theta_{k,j,t}\}$ corresponding to the J unselected options are set to 0, the estimated rewards $\{\hat{\theta}_{k,j,t}\}$ of the unselected options are 0 based on (7.20). Therefore, the values of $\{\omega_{k,j,t+1}\}$ corresponding to the J unselected options remain unchanged based on (7.21). Afterwards, SU_k updates $Q_k(t+1)$, $N_k(t+1)$, and $F_k(t+1)$ as (7.1) and (7.14). The iteration between decision making and learning continues until $t > T$.

C^2-EXP3 has the following advantages:

Context Awareness C^2-EXP3 achieves the context awareness through the dynamic adjustment of channel selection strategy based on the values of $F_k(t)$, $N_k(t)$, and $Q_k(t)$, which is similar to C^2-GSI.

Vigilant Attitude Towards Adversary During decision making, multiple SUs are prone to select the same channel with better empirical performance, thereby resulting severe co-channel interference and compromising the utility performance, i.e., channel selection strategies are adversarial. C^2-EXP3 holds a vigilant attitude towards adversary and can effectively resolve adversary contentions by continuing to explore the non-optimal options which may spontaneously perform well in future.

7.3   Performance Results and Discussions

We consider $K = 10$ SUs and $J = 5$ subchannels over a total optimization period of $T = 600$ slots. SUs are randomly distributed within a cell with a radius of 100 m. There are three licensed subchannels, i.e., $L = 3$, and two unlicensed subchannels in the 5 GHz band, i.e., $U = 2$. In the licensed spectrum, the bandwidth is $B_j = 10$ MHz and the noise power density is -17 dBm/Hz. In the unlicensed spectrum, the bandwidth is $B_j = 20$ MHz and the noise power density is -20 dBm/Hz [91]. The unit price of licensed subchannel is set as $r_j = 0.04$ cents/Mbit, $j = 1, 2, 3$. The other parameters are set as follows: $\tau = 1$ s, $\rho = 0.05$, $P_{TX} = 0.1$ W, $V_k = 7$, $\alpha_k = 100$, $\beta_k = 100$, $\eta_k = 0.7$, and $E_{k,max} = 25.8$ J, $\forall SU_k \in SU$. We consider that the SUs conduct video monitoring service where the videos captured have to be transmitted to the grid dispatch center. Taking the 1080P video as an example, the average data size of the generated video per second is 3.4 Mbits. Therefore, we assume $\varphi_k A_k(t)$ uniformly distributed in $[0.8\varphi_k \bar{A}_k(t), 1.2\varphi_k \bar{A}_k(t)]$ Mbits, where $\varphi_k \bar{A}_k(t) = 6.8$ Mbits is the time-average data size. $\varpi_{j,t}$ follows a uniform distribution within $[0.8\bar{\varpi}_{j,t}, 1.2\bar{\varpi}_{j,t}]$, where $\bar{\varpi}_{j,t} = 40\%$ represents the time-average duty cycle. The channel gain is defined as $g_{k,j,t} = 37 + 30\log_{10}(l_{k,t})$ [92], where $l_{k,t}$ is the distance between the SU_k and the LBS. The availability of licensed channel is shown in Table 7.1, which is determined by the activities of PUs, where $\checkmark$ represents that the channel is available, and × represents that the channel is unavailable. In comparison, unlicensed channels are available in all slots.

Four existing algorithms are employed for comparison. The first one is the conventional EXP3 algorithm without context awareness [93]. The second one is the conventional UCB, which is a classical algorithm for addressing stochastic MAB problems. It has been widely applied in resolving channel selection problems under information uncertainty [89, 94, 95]. Therefore, we select UCB as a comparison algorithm. The third one is C^2-UCB which augments UCB with context awareness, and the fourth one is random selection.

Figure 7.2 shows the cumulative utility performance. When $t = 600$, C^2-EXP3 outperforms EXP3, random selection, C^2-UCB, and UCB by 5.77%, 15.06%, 48.29%, and 59.91%, respectively. Both C^2-EXP3 and EXP3 outperform C^2-UCB and UCB dramatically due to the vigilant attitude towards adversary. UCB performs the worst because it inevitably leads multiple SUs to select the same channel

Table 7.1 Channel availability

Channel	c_1	c_2	c_3	c_4	c_5
Slot $1 \sim 200$	$\checkmark$	$\checkmark$	×	$\checkmark$	$\checkmark$
Slot $201 \sim 400$	$\checkmark$	×	$\checkmark$	$\checkmark$	$\checkmark$
Slot $401 \sim 600$	×	$\checkmark$	$\checkmark$	$\checkmark$	$\checkmark$

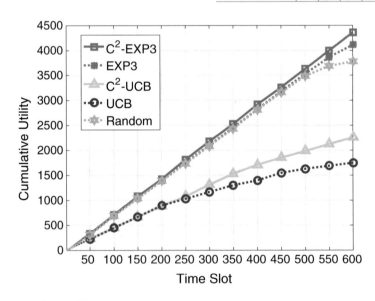

Fig. 7.2 Cumulative utility

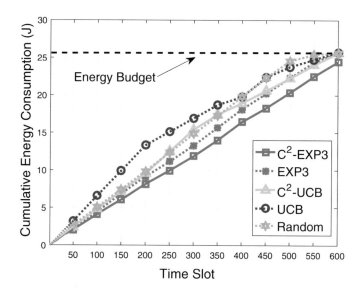

Fig. 7.3 Cumulative energy consumption

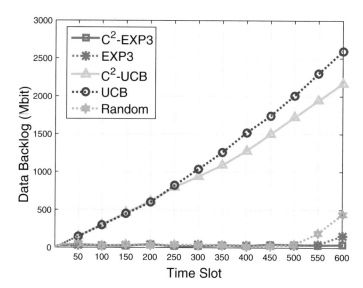

Fig. 7.4 Data backlog

with superior empirical performance, thereby resulting in severe interference and significant performance degradation.

Figure 7.3 shows the cumulative energy consumption versus time slot. Due to the energy awareness, the energy consumption of both C^2-EXP3 and C^2-UCB has not exceeded the energy budget. In comparison, random selection, EXP3, and UCB run out of energy around $t = 530$, $t = 570$, and $t = 590$ due to energy unawareness.

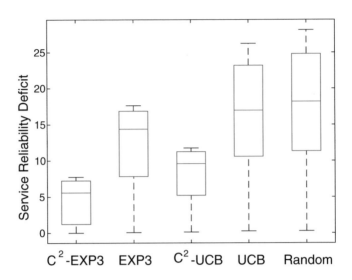

Fig. 7.5 Service reliability deficit

Figure 7.4 shows the data backlog versus time slot. Numerical results indicate that C^2-EXP3 can effectively reduce the data backlog. For instance, when $t = 600$, compared with EXP3, random selection, C^2-UCB, and UCB, C^2-EXP3 can reduce the data backlog by 82.58%, 93.68%, 98.75%, and 98.83%, respectively. The data backlogs of random selection and EXP3 grow dramatically after $t = 530$ and $t = 570$ due to the exhaustion of energy. UCB performs the worst because neither context awareness nor adversary has been taken into account.

Figure 7.5 shows the box plot of the service reliability deficit. The upper adjacent of the box is defined as the maximum data point located within the interval from the third quartile to the third quartile plus 1.5 times of interquartile range (IQR), and the lower adjacent is defined similarly. Simulation results demonstrate that the C^2-EXP3 can achieve the least service reliability deficit and performance fluctuation due to the service reliability awareness and the vigilant attitude towards adversary. However, the service reliability deficit of UCB is basically the same as that of random selection due to service reliability unawareness and slack attitude towards exploring potentially better channels.

Figure 7.6 shows the impact of the ratio of K/J on the average utility per SU. When the ratio is less than 1, both C^2-EXP3 and EXP3 perform worse than C^2-UCB and UCB due to the overestimation of adversary. As the ratio increases, the adversary among SUs is gradually enhanced. C^2-UCB and UCB cannot handle the strong adversary and their average utilities degrade gradually as the number of SUs increases. In comparison, C^2-EXP3 and EXP3 perform better because adversary has been taken into consideration. Specifically, compared with C^2-UCB, C^2-EXP3 can improve the average utility by 48.29% when $K/J = 2$. However, when the ratio is too large, i.e., $K/J > 2$, the channel selection competition among SUs becomes so intense, and the average utilities of C^2-EXP3 and EXP3 also degrade due to

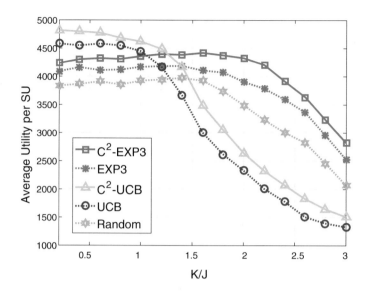

Fig. 7.6 Average utility versus K/J

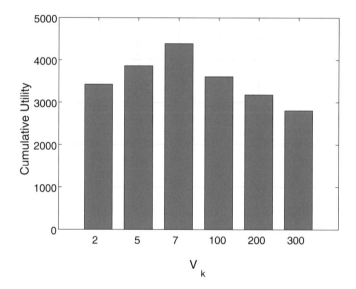

Fig. 7.7 Impact of V_k

the strong interference and poor subchannel availability. Nevertheless, simulation results demonstrate that the C^2-EXP3 achieves the highest average utility due to the consideration of context awareness.

Figure 7.7 shows the impact of parameter V_k on utility. By increasing V_k from 2 to 300, the cumulative utility increases firstly and then decreases. A larger V_k

indicates less concern on energy budget and service reliability, thereby the impacts of service reliability awareness and energy awareness are weakened. A smaller V_k indicates more concern on energy budget which results in a conservative attitude towards data transmission. These results indicate that the parameter V_k should be chosen appropriately to maintain the tradeoff between utility and other performance metrics.

In this chapter, we demonstrated a context-aware channel selection algorithm based on EXP3 named C^2-EXP3 for edge computing-empowered CM2M communications with the coexistence of licensed and unlicensed spectrum. Simulation results demonstrate that the C^2-EXP3 can improve the utility by 5.77%, 15.06%, 48.29%, and 59.91%, compared with EXP3, random selection, C^2-UCB, and UCB while stabilizing the data queue and satisfying the constraints of energy consumption and service reliability.

Chapter 8
Energy-Efficient Task Assignment and Route Planning for UAV

8.1 Framework of UAV-Aided MCS Systems

In this chapter, we demonstrate a UAV sensing scenario which consists of a target sensing region, a MCS carrier, and several UAVs embedded with various sensors and communication devices. A conceptual illustration of the system model is shown in Fig. 8.1. The target sensing region, which includes a number of target points to be sensed, is divided into N smaller individual subregions due to the following benefits: first, considering the limited battery capacity, the area of the whole sensing region is generally beyond the maximum flying range of any single UAV; second, the sensing tasks in different subregions can be performed simultaneously by multiple UAVs to reduce the overall sensing latency. We have adopted fixed-wing UAVs for task sensing. Compared to a multi-rotor UAV, a fixed-wing UAV is able to fly longer distances and cover much larger areas, since most of the energy has been consumed for moving forward rather than holding the UAV up in the air [96]. Nevertheless, the algorithm can also be extended to the cases of multi-rotor UAVs by modifying the energy consumption model.

Denoting the set of subregions as $\mathcal{R}_N = \{R_1, \cdots, R_n, \cdots, R_N\}$, the topology of any subregion $R_n \in \mathcal{R}_N$ can be modeled as a graph. Let $\mathcal{I}_n = \{1, \ldots, i, \ldots, I_n\}$ denote the set of nodes in the graph representation of subregion R_n, the elements of which are also the sensing target points. The route segment between target point i and i', i.e., $\forall i, i' \in \mathcal{I}_n, i \neq i'$, is denoted as $l_{i,i'}^n$, and the length of segment $l_{i,i'}^n$ is defined as $d_{i,i'}^n$.

Remark 8.1 (Sensing Region Partition) The sensing region can be transformed into a graph $\mathcal{G}_0 = \{\mathcal{I}_0, \mathcal{L}_0\}$, where $\mathcal{I}_0$ is the set of vertices, and $\mathcal{L}_0$ is the set of edges. An N-way graph partition problem is formulated to partition $\mathcal{G}_0$ into several smaller graphs [97], where the number of vertices in each smaller graph is roughly equal so that the sensing task is balanced among N subregions. The N-way graph partition

© Springer Nature Switzerland AG 2021

Z. Zhou et al., *Green Internet of Things (IoT): Energy Efficiency Perspective*,
Wireless Networks, https://doi.org/10.1007/978-3-030-64054-5_8

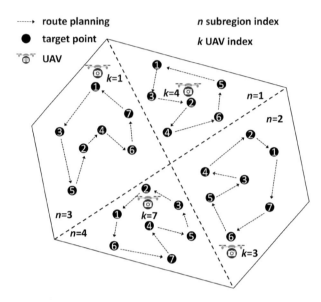

Fig. 8.1 The system model of UAV-aided MCS systems

problem can be effectively solved by the multilevel graph partition algorithm [97], which consists of the following three phases:

- **Coarsening phase:** Transform the original graph $\mathcal{G}_0$ into a number of smaller graphs $\mathcal{G}_1, \mathcal{G}_2, \cdots, \mathcal{G}_m$ such that $|\mathcal{I}_0| > |\mathcal{I}_1| > |\mathcal{I}_2| \cdots > |\mathcal{I}_m|$. Here, $|\mathcal{I}_m|$ denotes the number of vertices in graph $\mathcal{G}_m$.
- **Partitioning phase:** An N-way partition $\mathcal{Z}_N$ is utilized to partition the coarser graph $\mathcal{G}_m$ into N parts, where each part contains approximate $1/N$ of the vertices.
- **Uncoarsening phase:** Project the N-way partition $\mathcal{Z}_N$ of $\mathcal{G}_m$ back to the original graph $\mathcal{G}_0$ by going through all the intermediate partitions $\mathcal{Z}_{m-1}, \mathcal{Z}_{m-2}, \ldots, \mathcal{Z}_1, \mathcal{Z}_0$.

The joint optimization of sensing region partition, route planning, and task assignment for UAV-aided MCS systems is still an open problem, which is out of the scope of this work and will be investigated in the future.

Remark 8.2 If there exists no blockage, then any pair of two target points in subregion R_n, e.g., i and i', are directly connected. Hence, $d_{i,i'}^n$ represents the Euclidean distance between i and i' and $d_{i,i'}^n < \infty$. Moreover, our work can also be extended to a more general scenario where there exists no path between i and i' due to the blockage or other constraints such as government regulations. In this case, i and i' are not interconnected and $d_{i,i'}^n = \infty$.

The MCS process is implemented as follows. At the beginning, the MCS carrier issues sensing requests (including the information of sensing subregions) and recruits UAVs to perform sensing tasks via an Internet-based sensing platform.

The set of available UAVs is denoted as $\mathcal{U}_K = \{U_1, \cdots, U_k, \cdots, U_K\}$, where K represents the total number of UAVs. Then, each UAV performs the sensing task based on the task assignment decisions. Assuming that subregion R_n is assigned, the UAV starts from target point $i = 0$, and flies along segments to pass through every target point in subregion $\mathcal{I}_n$. Upon finishing the sensing task, the UAV returns back to the initial position, and uploads the collected data to the platform via Internet. We assume that the initial and final location of each UAV is pre-determined since in practical world, a UAV is launched by its owner, whose location is also pre-determined. Nevertheless, the initial and final locations of different UAVs are independent, which does not impact the solution structure of the algorithm.

Remark 8.3 When the number of UAVs is sufficient, it is intuitive to partition the target sensing region into N ($N \leq K$) smaller subregions to enable parallel sensing. If $K < N$, the algorithm can also be applied. However, some subregions will be left unmatched due to the limited number of available UAVs, the impact of which is investigated via simulations.

The task assignment decisions are defined as follows.

Definition 8.1 (Task Assignment Decision) The task assignment decisions of the MCS carrier and UAVs are denoted as a $N \times K$ matrix $\mathbf{X}_{N \times K}$ and a $K \times N$ matrix $\mathbf{Y}_{K \times N}$, respectively. Here, the (n, k)-th element of $\mathbf{X}_{N \times K}$, i.e., $x_{n,k}$, is a binary value, where $x_{n,k} = 1$ represents that the MCS carrier prefers UAV U_k to perform the sensing task of subregion R_n, and otherwise, $x_{n,k} = 0$. Similarly, the (k, n)-th element of $\mathbf{Y}_{K \times N}$, i.e., $y_{k,n}$, is also a binary value, where $y_{k,n} = 1$ represents that UAV U_k prefers to sense subregion R_n, and otherwise, $y_{k,n} = 0$.

Remark 8.4 Since the MCS carrier and UAVs are owned by different entities, we assume that any MCS participant involved is selfish, and only concerns about its own benefit. Hence, it is very likely that the MCS carrier and UAVs have completely conflicting task assignment decisions, e.g., $x_{n,k} \neq y_{k,n}$. A sensing task can be implemented if and only if a mutual agreement on task assignment has been reached, i.e., $x_{n,k} = y_{k,n} = 1$.

In the following, we elaborate the utility models of the MCS carrier and UAVs.

8.1.1 The Utility Function of the MCS Carrier

For the MCS carrier, its revenue depends on how well and how fast the sensing tasks can be finished, i.e., sensing quality, quantity, and latency. Hence, the revenue of the MCS carrier, which is achieved by assigning UAV U_k to sense subregion R_n, is calculated as

$$g_{n,k} = \alpha_n D_n \delta_k t_{k,n}^{-1}, \tag{8.1}$$

where α_n is the price coefficient, which is used to transform various key parameters such as sensing quality, quantity, and latency, into a unified unit, e.g., dollars. D_n is the total amount of data that has to be collected in subregion R_n, which is positively correlated with the number of target points I_n. δ_k is the quality of the data collected by UAV U_k, which is related to the accuracy of the sensing equipment. For the sake of simplicity, both D_n and δ_k are defined as fixed parameters for subregion R_n and UAV U_k, respectively. $t_{k,n}$ is the total duration needed by UAV U_k for finishing the sensing task in subregion R_n, which depends on both the topology of subregion R_n and the route planning decision of UAV U_k. Here, $t_{k,n}^{-1}$ is just utilized as an example to show that $g_{n,k}$ is inversely proportional to the sensing latency. The system model and solutions can be easily extended to more complicated expressions.

For each assigned task, the MCS offers a payment to the recruited UAV based on the quality and quantity of collected data. The payment offered to UAV U_k for sensing subregion R_n is expressed as

$$p_{n,k} = \beta_n D_n \delta_k, \tag{8.2}$$

where β_n is the price coefficient which plays a similar role as α_n.

Hence, the payoff of the MCS carrier in subregion R_n is calculated as the total revenue minus the payments offered to UAVs, which is expressed as

$$G_n(\mathbf{x}_n) = \sum_{k=1}^{K} x_{n,k} y_{k,n} (g_{n,k} - p_{n,k}), \tag{8.3}$$

where $\mathbf{x}_n$ is the n-th row of the matrix $\mathbf{X}_{N \times K}$, i.e., $\mathbf{x}_n = \{x_{n,1}, \cdots, x_{n,k}, \cdots, x_{n,K}\}$.

8.1.2 The Utility Function of UAVs

For UAV U_k, the reward gained from the MCS carrier for sensing subregion R_n is exactly $p_{n,k}$ defined in (8.3). The corresponding cost is represented as a linear function of the total energy consumption, which consists of the energy consumed for propulsion, direction adjustment, data collection and transmission. Generally, the energy consumption of propulsion and direction adjustment is several orders of magnitudes higher than that of the communication, where the latter can be ignored to reduce the computation complexity [98].

For fixed-wing UAVs, the energy consumption of propulsion can be derived based on the steady straight-and-level flight (SSLF) model proposed in [99], which indicates the following two assumptions: (1) any UAV flies towards a dedicated direction in a constant speed without horizontal acceleration and sudden turning; (2) any UAV flies at a constant altitude without vertical acceleration because of the lift-weight balance. Therefore, it is assumed that the velocity of each UAV remains constant during a segment, and may change from segment to segment. Considering a scenario where UAV U_k flies with a constant velocity v, the required propulsion power is given by Zeng and Zhang [99]

$$P_k^p(v) = c_{k,1}v^3 + \frac{c_{k,2}}{v}, \tag{8.4}$$

where $c_{k,1}v^3$ is the required power to balance the parasitic drag caused by skin friction, and $c_{k,2}/v$ is the required power to balance the drag force of air redirection. $c_{k,1}$ and $c_{k,2}$ are related to numerous parameters about the UAV and the environment, which are calculated as

$$c_{k,1} \triangleq \frac{1}{2}\rho C S_k,$$

$$c_{k,2} \triangleq \frac{2W_k^2}{(\pi e\eta_k)\rho S_k}, \tag{8.5}$$

where ρ and C represent the air density and the coefficient of zero-lift drag force, respectively. S_k is the reference area of UAV U_k. e is the Osawald efficiency. η_k and W_k are the wing aspect ratio and the weight of UAV U_k, respectively.

Thus, the amount of energy consumed by UAV U_k for flying a distance d with a constant velocity v is given by

$$E_k^p(v) = \frac{d}{v}P_k^p(v) = d\frac{c_{k,1}v^4 + c_{k,2}}{v^2}. \tag{8.6}$$

If UAV U_k changes its heading direction, the corresponding energy consumption is given by

$$E_k^h(v) = \frac{c_{k,2}}{vg^2}\int_0^{T_{k'}} a^2(t)dt, \tag{8.7}$$

where $a(t)$ represents the acceleration of UAV U_k during head direction adjustment. $T_{k'}$ represents the total adjustment duration, and g is the gravitational acceleration.

For example, we assume that $g = 9.8$ m/s^2, $T_{k'} = 1$ s, $a(t) = 5$ m/s^2, $d = 10^3$ m and $v = 6.69$ m/s. Then, we can calculate that $E_k^p = 8.944 \times 10^3$ J and $E_k^h = 7.8$ J. Hence, the energy consumption of propulsion is more than 10^3 times higher than that of head direction adjustment, i.e., $E_k^p \gg E_k^h$, which can be ignored without obvious performance degradation. This assumption holds true when the UAV does not adjust direction frequently during a short-distance flight, which is valid for most task-sensing scenarios since there usually exists a straight-line path between two adjacent target points. The more complicated scenario where the energy consumption of head direction is comparable to or even higher than that of propulsion is out of the scope of this chapter and will be studied in the future work.

For UAV U_k, the velocity at each segment should be dynamically adjusted in order to reduce energy consumption. We use $v_{i,i'}^{k,n}$ to denote the velocity of UAV U_k when flying through segment $l_{i,i'}^n$ of subregion R_n. The required propulsion power $P_{k,l_{i,i'}^n}^p$ and the corresponding propulsion energy consumption $E_{k,l_{i,i'}^n}^p$ can be calculated based on (8.4) and (8.6) accordingly by replacing v and d with $v_{i,i'}^{k,n}$ and $d_{i,i'}^n$, respectively. Throughout this work, the air density is assumed as the same within the sensing region to simplify the expression of the propulsion power as well as the energy consumption.

The set of segment selection strategies for subregion R_n is defined as $S_{k,n} = \{s_{i,i'}^{k,n} \mid \forall i, i' \in \mathcal{I}_n, i \neq i'\}$, and the set of velocity control strategies is defined as $\mathcal{V}_{k,n} = \{v_{i,i'}^{k,n} \mid \forall i, i' \in \mathcal{I}_n, i \neq i'\}$. $s_{i,i'}^{k,n}$ is the selection decision for segment $l_{i,i'}^n$, i.e., $s_{i,i'}^{k,n} = 1$ represents that UAV U_k will fly through segment $l_{i,i'}^n$, and $s_{i,i'}^{k,n} = 0$, otherwise.

The time required for UAV U_k to fly from i to i' is calculated as

$$t_{i,i'}^{k,n}(v_{i,i'}^{k,n}) = \frac{d_{i,i'}^n}{v_{i,i'}^{k,n}}. \tag{8.8}$$

The total sensing latency $t_{k,n}$ is defined as the total flying time for UAV U_k to fly through all the target points in subregion R_n, which is calculated as

$$t_{k,n}(S_{k,n}, \mathcal{V}_{k,n}) = \sum_{i \in \mathcal{I}_n} \sum_{i' \neq i, i' \in \mathcal{I}_n} s_{i,i'}^{k,n} t_{i,i'}^{k,n}(v_{i,i'}^{k,n}). \tag{8.9}$$

The total cost of UAV U_k for sensing subregion R_n is defined as a linear function of the total energy consumption, which is calculated as

$$c_{k,n}(S_{k,n}, \mathcal{V}_{k,n}) = \sigma_{k,n} \sum_{i \in \mathcal{I}_n} \sum_{i' \neq i, i' \in \mathcal{I}_n} s_{i,i'}^{k,n} E_{k,l_{i,i'}^n}^P (v_{i,i'}^{k,n}), \tag{8.10}$$

where $\sigma_{k,n}$ is the price coefficient. Thus, the profit of UAV U_k is calculated as

$$G_k(\mathbf{y}_k) = \sum_{n=1}^N x_{n,k} y_{k,n} (p_{n,k} - c_{k,n}), \tag{8.11}$$

where $\mathbf{y}_k$ is the k-th row of the matrix $\mathbf{Y}_{K \times N}$, and $\mathbf{y}_k = \{y_{k,1}, \cdots, y_{k,n}, \cdots, y_{k,N}\}$.

Remark 8.5 The roles of price coefficients such as α_n, β_n, and $\sigma_{k,n}$ are three folds: (1) to transform various factors with different units into a unified unit to characterize their impacts on profits; (2) to adjust the price coefficients to guarantee individual rationality for both the MCS carrier and UAVs; (3) to dynamically adjust the priority of certain subregion. For example, subregion R_n will get a higher priority by increasing α_n and decreasing $\sigma_{k,n}$.

8.1.3 UAV-Aided MCS Systems Problem Formulation

The objective of the MCS carrier is to maximize its profit defined in (8.3) under the constraint of sensing latency. The problem formulation for the MCS carrier is given by

$$\textbf{P1}: \max_{\mathbf{X}_{N \times K}} \sum_{n=1}^{N} G_n(\mathbf{x}_n)$$

$$\text{s.t.} \quad C_1 : t_{k,n} \leq T_n^{max}, \forall R_n \in \mathcal{R}_N, \forall U_k \in \mathcal{U}_K,$$

$$C_2 : x_{n,k} = \{0, 1\}, \forall R_n \in \mathcal{R}_N, \forall U_k \in \mathcal{U}_K,$$

$$C_3 : \sum_{n=1}^{N} x_{n,k} \leq 1, \forall U_k \in \mathcal{U}_K,$$

$$\sum_{k=1}^{K} x_{n,k} \leq 1, \forall R_n \in \mathcal{R}_N. \tag{8.12}$$

Here, C_1 represents that the time required to sense any subregion $R_n \in \mathcal{R}_N$ should be less than the specified latency threshold T_n^{max}. C_2 and C_3 guarantee that each UAV is allowed to sense at most one subregion, and each subregion is sensed by at most one UAV.

The objective of each UAV $U_k \in \mathcal{U}_K$ is also to maximize its profit defined in (8.11) under the constraint of battery capacity. The problem formulation for UAV U_k is given by

$$\textbf{P2}: \max_{\mathbf{y}_k, \mathcal{S}_{k,n}, \mathcal{V}_{k,n}, \forall R_n \in \mathcal{R}_N} G_k(y_{k,n})$$

$$\text{s.t.} \quad C_4 : \sum_{i \in \mathcal{I}_n} \sum_{i' \neq i, i' \in \mathcal{I}_n} s_{i,i'}^{k,n} E_{k,l_{i,i'}^n}^{p}(v_{i,i'}^{k,n}) \leq E_k^{max},$$

$$\forall R_n \in \mathcal{R}_N,$$

$$C_5 : \sum_{i \in \mathcal{I}_n} \sum_{i' \neq i, i' \in \mathcal{I}_n} s_{i,i'}^{k,n} = I_n - 1, \forall R_n \in \mathcal{R}_N,$$

$$C_6 : v_{i,i',min}^{k,n} \leq v_{i,i'}^{k,n} \leq v_{i,i',max}^{k,n},$$

$$\forall i, i' \in \mathcal{I}_n, i \neq i', \forall R_n \in \mathcal{R}_N,$$

$$C_7 : y_{k,n} = \{0, 1\}, \forall R_n \in \mathcal{R}_N, \forall U_k \in \mathcal{U}_K,$$

$$C_8 : \sum_{n=1}^{N} y_{k,n} \leq 1, \forall R_n \in \mathcal{R}_N,$$

$$\sum_{k=1}^{K} y_{k,n} \leq 1, \forall U_k \in \mathcal{U}_K. \tag{8.13}$$

Here, C_4 is the energy consumption constraint, i.e., the total amount of energy consumed for propulsion should be less than the battery capacity. C_5 is the constraint of miss detection, which makes sure that all the target points in any subregion $R_n \in \mathcal{R}_N$ will be sensed. C_6 represents the boundary constraint of velocity for each segment. C_7 and C_8 act similar roles as C_2 and C_3.

Remark 8.6 Considering subregion R_n and UAV U_k, the route planning decision of UAV U_k, which includes segment selection and velocity control, not only determines the total amount of energy consumed for propulsion, but also determines the total sensing cost $c_{k,n}$ and the total sensing latency $t_{k,n}$. The route planning problem is coupled with the task assignment problem since the profit of the MCS carrier in subregion R_n, i.e., G_n, and the profit of UAV U_k, i.e., G_k, depend on $t_{k,n}$ and $c_{k,n}$, respectively.

8.2 Energy-Efficient Joint Task Assignment and Route Planning

In this section, we introduce the energy-efficient joint task assignment and route planning algorithm.

8.2.1 Problem Transformation

As analyzed in Sect. 8.1.3, it is infeasible to solve **P1** and **P2** in polynomial time since the task assignment and route planning problems are coupled with each other. Particularly, the optimal task assignment strategies of both the MCS carrier and UAVs rely on the sensing cost and sensing latency of each UAV, which have to be decided initially by solving the route planning problem.

To provide a tractable solution, the original NP-hard problem can be transformed into a two-stage two-sided matching problem based on the problem structure and chronological order of decisions [57]. The route planning problem is solved in the first stage to obtain the preferences of UAVs toward subregions, and vice versa. The task assignment problem is solved in the second stage to derive the matching between two sides, i.e., N subregions on one side, and K UAVs on the other side.

The transformed matching problem can be formulated as a triple $(\mathcal{R}_N, \mathcal{U}_K, \mathcal{F})$, in which $\mathcal{R}_N$ and $\mathcal{U}_K$ are two distinct and finite sets of matching participants, i.e., subregions and UAVs, respectively. $\mathcal{F}$ represents the set of mutual matching preferences. Each participant, either a subregion or a UAV, aims at maximizing its individual profit under the specified constraints. A one-to-one matching ϕ is defined as [32]:

Definition 8.2 (Matching) For the formulated matching problem $(\mathcal{R}_N, \mathcal{U}_K, \mathcal{F})$, the matching ϕ is a one-to-one correspondence from set $\mathcal{R}_N \cup \mathcal{U}_K$ onto itself under preference $\mathcal{F}$, i.e., $\phi(R_n) \in \mathcal{U}_K \cup \{R_n\}$ and $\phi(U_k) \in \mathcal{R}_N \cup \{U_k\}$, $\forall R_n \in \mathcal{R}_N$, $\forall U_k \in \mathcal{U}_K$. $\phi(R_n) = U_k$ if and only if $\phi(U_k) = R_n$. Here, if $\phi(R_n) = R_n$ or $\phi(U_k) = U_k$, R_n or U_k remains unmatched.

8.2.2 The Route Planning

(1) DP Based First-Stage Route Planning

In the first stage, the route planning problem has to be solved in order to obtain the minimum energy consumption, which is essential to derive the preferences of UAVs. The energy consumption of UAV U_k to perform the sensing tasks in subregion R_n is minimized by solving the following route planning problem

$$\mathbf{P3}: \min_{\mathcal{S}_{k,n}, \mathcal{V}_{k,n}} \sum_{i \in \mathcal{I}_n} \sum_{i' \neq i, i' \in \mathcal{I}_n} s_{i,i'}^{k,n} E_{k,l_{i,i'}^n}^P (v_{i,i'}^{k,n})$$

$$\text{s.t.} \quad C_4, C_5. \tag{8.14}$$

It is noted that the route planning problem **P3** has two unique features:

- The route planning decisions are made dynamically in discrete-time stages.
- The energy consumption is additive over stages, i.e., the total energy consumption is the sum of the propulsion energy consumed for flying through each selected segment.

Hence, **P3** falls within the framework of the deterministic finite-state travel salesman problem (TSP), which is a typical NP hard.

The principle of DP is that the system state evolves in accordance with the decisions that are made in discrete stages. The system state evolution is given by

$$r_{\tau+1} = z(r_\tau, u_\tau), \tau = 0, 1, 2, \cdots, \psi - 1, \tag{8.15}$$

where ψ and τ represent the total number of stages and the index of stage, respectively. r_τ and $r_{\tau+1}$ represent the system states at stage τ and stage $\tau + 1$, respectively. u_τ is the decision taken at stage τ. The state updating mechanism is enumerated by z. To make the DP formulation consistent with the route planning problem, we denote the set of states as the set of target points. For example, if the state at stage τ is target point i, then $r_\tau = i$, $\forall i \in \mathcal{I}_n$. The set of decisions is the same as $\mathcal{S}_{k,n}$. If segment $l_{i,i'}^n$ is selected at stage τ, then $u_\tau = s_{i,i'}^{k,n}[\tau]$. Hence, the state is updated as

$$r_{\tau+1} \mid_{r_\tau = i} = \sum_{i' \neq i, i' \in I_n} i' s_{i,i'}^{k,n}[\tau]. \tag{8.16}$$

The cost function of DP at stage τ, i.e., the amount of energy consumed at stage τ, is calculated as

$$C^o(r_\tau, u_\tau) \mid_{(r_\tau = i, u_\tau = s_{i,i'}^{k,n}[\tau])} = \sum_{i' \neq i, i' \in I_n} s_{i,i'}^{k,n}[\tau] E_{k,l_{i,i'}^n}^P (v_{i,i'}^{k,n*}). \tag{8.17}$$

Here, $v_{i,i'}^{k,n*}$ is the optimal velocity for segment $l_{i,i'}^n$, and $E_{k,l_{i,i'}^n}^P (v_{i,i'}^{k,n*})$ is the minimum amount of energy consumed by the UAV for flying from target point i to i' in subregion R_n. The energy consumption is additive over stage. Assuming that the initial stage as $r_0 = 0$ ($\tau = 0, i = 0$) and ψ is finite, i.e., the number of target points is limited, the minimum energy consumption of UAV U_k for sensing subregion R_n is defined as

$$J^*(f_0) = \min_{S_{k,n}} \left[\sum_{\tau=0}^{\psi-1} C^o(r_\tau, u_\tau) + C^o(\psi) \right], \tag{8.18}$$

where $C^o(\psi)$ represents the minimum energy consumption at the last stage, i.e., the terminal cost. The above optimization problem can be solved by proceeding segment selection backward from stage $\psi - 1$ to stage 0 as

$$J_\psi(r_\psi) = C^o(r_\psi), \tag{8.19}$$

$$J_\tau(r_\tau) = \min_{s_{i,i'}^{k,n}(\tau)} \{ C^o(r_\tau, u_\tau) + J_{\tau+1}(z(r_\tau, u_\tau)) \},$$

$$\tau = 0, 1, \cdots, \psi - 1. \tag{8.20}$$

However, it is observed that the minimum energy consumption at any segment, e.g., $E_{k,l_{i,i'}^n}^P (v_{i,i'}^{k,n*})$, $i' \neq i, i, i' \in I_n$, is still unknown. To obtain $E_{k,l_{i,i'}^n}^P (v_{i,i'}^{k,n*})$, the following velocity control problem needs to be solved.

$$\mathbf{P4} : \min_{v_{i,i'}^{k,n}} E_{k,l_{i,i'}^n}^P (v_{i,i'}^{k,n}) = d_{i,i'}^n \frac{c_{i,i'}^{1,n} (v_{i,i'}^{k,n})^4 + c_{i,i'}^{2,n}}{(v_{i,i'}^{k,n})^2}$$

s.t. C_6. \tag{8.21}

It is noted that **P4** is convex, and can be solved by using Karush-Kuhn-Tucker (KKT) conditions. The associated Lagrangian of **P4** is given as

$$L(v_{i,i'}^{k,n}, \lambda, \mu) = E_{k,l_{i,i'}^n}^p(v_{i,i'}^{k,n}) + \lambda(v_{i,i'}^{k,n} - v_{i,i',max}^{k,n})$$

$$-\mu(v_{i,i',min}^{k,n} - v_{i,i'}^{k,n}), \tag{8.22}$$

where μ and λ are the Lagrange multipliers. The optimal velocity can be obtained by taking the first-order derivative of $L(v_{i,i'}^{k,n}, \lambda, \mu)$ with respect to $v_{i,i'}^{k,n}$, λ and μ, respectively, which is calculated by solving the following functions:

$$
\begin{cases}
\dfrac{\partial L(v_{i,i'}^{k,n}, \lambda, \mu)}{\partial v_{i,i'}^{k,n}} = 0, \\[4mm]
\dfrac{\partial L(v_{i,i'}^{k,n}, \lambda, \mu)}{\partial \lambda} = 0, \\[4mm]
\dfrac{\partial L(v_{i,i'}^{k,n}, \lambda, \mu)}{\partial \mu} = 0.
\end{cases}
\tag{8.23}
$$

The total sensing latency and sensing cost are obtained as

$$t_{k,n}^* = \sum_{i \in I_n} \sum_{i' \neq i, i' \in I_n} s_{i,i'}^{k,n} \frac{d_{i,i'}^n}{v_{i,i'}^{k,n*}}, \tag{8.24}$$

$$c_{k,n}^* = \sigma_n \sum_{i \in I_n} \sum_{l' \neq l, l' \in I_n} s_{i,i'}^{k,n} E_{k,l_{i,i'}^n}^p(v_{i,i'}^{k,n*}). \tag{8.25}$$

(2) GA Based Low-Complexity Route Planning
The computation complexity of DP becomes prohibitive as the problem size grows. Therefore, we propose a low-complexity suboptimal route planning algorithm based on GA. GA has been widely used for solving combinatorial optimization problems [100]. The GA-based route planning algorithm consists of the following several components:

- **Initialization:** Given subregion R_n and UAV U_k, each possible route planning solution is expressed in the form of a string or chromosome. The set of chromosomes is defined as $Q_k = \{r_0, \cdots, r_{\tau-1}, r_\tau, \cdots, r_\psi\}$, $\forall r_\tau \in I_n$, $r_\tau \neq r_{\tau-1}$, $\tau = 1, 2, \cdots, \psi$. Here, the elements within each chromosome are called genes. The initial population is generated by randomly selecting the value of each gene from the set I_n, e.g., $r_\tau = i$.
- **Fitness and unfitness evaluation:** For any chromosome $q \in Q_k$, we can derive the corresponding set of segment selection strategies $S_{k,n}(q)$ and the set of velocity control strategies $V_{k,n}(q)$. For example, if $r_\tau = i$ and $r_{\tau+1} = i'$, we

have $s_{i,i'}^{k,n}[\tau] = 1$. The fitness level of q is calculated as the reciprocal of the total sensing cost, which is calculated as $f(q) = 1/\big(c_{k,n}[\mathcal{S}_{k,n}(q), \mathcal{V}_{k,n}(q)]\big)$.

- **Crossover and mutation:** By choosing two parent chromosomes q_1 and q_2, the partially matched crossover (PMX) is employed to reproduce two offspring chromosomes $\hat{q}_1$ and $\hat{q}_2$. An example is shown in Fig. 8.2. Specifically, we randomly choose two points in the parent chromosomes, and replace the genes of q_1 between these two points, i.e., 3, 4, 5, 6, by using the corresponding genes of q_2, i.e., 6, 9, 2, 1, and vice versa. After the initial crossover, a conflict may occur if a chromosome contains duplicate copies of the same gene, e.g., 1, 2, 9 in q_1. Then, the duplicated genes 1, 2, 9 are replaced by 3, 5, 4 based on a one-to-one mapping, i.e., $1 \rightarrow 6 \rightarrow 3$, $2 \rightarrow 5$, and $9 \rightarrow 4$. Next, the mutation procedure is implemented, in which the genes between two randomly chosen points are reversed.
- **Replacement:** Inferior offspring chromosomes have to be removed to maintain the population size as a constant. We adopt the ranking replacement method [101], which replaces parent chromosome q_1 by its offspring $\hat{q}_1$ if $f(\hat{q}_1) > f(q_1)$.

After population initialization, the procedures of crossover, mutation, and replacement are repeated until a termination criterion is reached. Various criteria can be utilized such as the elapsed computation time, the number of generations, and the performance improvement threshold, etc. Analogous to biological evolution, offspring solutions with better fitness have a higher probability to survive and reproduce, and the fitness level of their offsprings is expected to be improved as the population evolves.

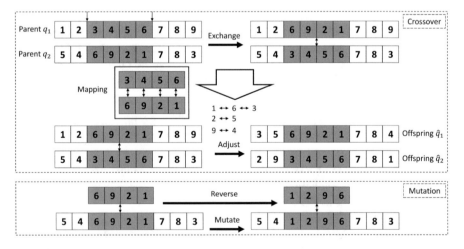

Fig. 8.2 The process of crossover and mutation

8.2.3 Preference List Construction

In order to implement the two-sided matching, each matching participant needs to construct its preference list by ranking participants from the other side in accordance with preference. Algorithm 8.1 provides a summary about how to construct preference lists based on route planning results. To achieve profit maximization, we model the preference as the maximum profit that can be achieved. For example, the preference of subregion R_n towards UAV U_k, or vice versa, is defined as the maximum profit that is achieved under the matching $\phi(R_n) = U_k$, i.e., ($x_{n,k} = y_{k,n} = 1$). The mutual preference of subregion R_n and UAV U_k are calculated as

$$G_n^*\big|_{\phi(R_n)=U_k} = g_{n,k} - p_{n,k}, \tag{8.26}$$

$$G_k^*\big|_{\phi(U_k)=R_n} = p_{n,k} - c_{k,n}. \tag{8.27}$$

We introduce a complete, reflexive, and transitive binary preference relation [32], i.e., "$\succ$", to compare the preferences. For instance, $R_n \succ_{U_k} R_n'$ represents that UAV U_k prefers subregion R_n to subregion R_n', which is given by

$$R_n \succ_{U_k} R_n' \Leftrightarrow G_k^*\big|_{\phi(U_k)=R_n} > G_k^*\big|_{\phi(U_k)=R_n'}. \tag{8.28}$$

Furthermore, $R_n \succeq_{U_k} R_n'$ represents that UAV U_k prefers subregion R_n at least as well as subregion R_n', which is given by

$$R_n \succeq_{U_k} R_n' \Leftrightarrow G_k^*\big|_{\phi(U_k)=R_n} \geq G_k^*\big|_{\phi(U_k)=R_n'}. \tag{8.29}$$

Similarly, $U_k \succ_{R_n} U_k'$ represents that subregion R_n prefers UAV U_k to UAV U_k', which is expressed as

$$U_k \succ_{R_n} U_k' \Leftrightarrow G_n^*\big|_{\phi(R_n)=U_k} > G_n^*\big|_{\phi(R_n)=U_k'}. \tag{8.30}$$

We use F^n and F^k to denote the preference list of subregion R_n towards all the UAVs, and the preference list of UAV U_k towards all the subregions, respectively, $\forall R_n \in \mathcal{R}_N, \forall U_k \in \mathcal{U}_K$. F^n is obtained by sorting all the K UAVs in a descending order according to $G_n^*\big|_{\phi(R_n)=U_k}, \forall U_k \in \mathcal{U}_K$. Similarly, the preference list F^k of UAV U_k is obtained by sorting all of N subregions according to the obtained $G_k^*\big|_{\phi(U_k)=R_n}, \forall R_n \in \mathcal{R}_N$. The total set $\mathcal{F}$ is constructed as $\mathcal{F} = \{F^n, F^k \mid \forall R_n \in \mathcal{R}_N, \forall U_k \in \mathcal{U}_K\}$.

Algorithm 8.1 Preference list establishment with DP-based route planning

---

1: **Input:** $\mathcal{R}_N, \mathcal{U}_K$.
2: **Output:** $\mathcal{F}$.
3: **for** $R_n \in \mathcal{R}_N$ **do**
4:    **for** $U_k \in \mathcal{U}_K$ **do**
5:       $\tau = I_n$;
6:       $\Omega = \mathcal{I}_n$.
7:       **while** $\Omega \neq \emptyset$ **do**
8:          **for** $i \in \Omega$ **do**
9:             calculate $C^o(r_\tau, u_\tau) \big|_{(r_\tau = i, u_\tau = s_{i,i'}^{k,n}[\tau])}$;
10:         **end for**
11:         Calculate $J_\tau(r_\tau)$ based on (8.20), and obtain the optimal route plan, set $s_{i,i'}^{k,n} = 1$;
12:         update i' as i;
13:         remove i from Ω;
14:         $\tau = \tau - 1$.
15:      **end while**
16:      calculate $G_n^* \big|_{\phi(R_n)=U_k}$ and $G_k^* \big|_{\phi(U_k)=R_n}$ based on (8.26) and (8.27).
17:   **end for**
18: **end for**
19: **for** $R_n \in \mathcal{R}_N$ **do**
20:    sort every $U_k \in \mathcal{U}_K$ in a descending order based on the obtained $G_n^* \big|_{\phi(R_n)=U_k}$ to form F^n.
21: **end for**
22: **for** $U_k \in \mathcal{U}_K$ **do**
23:    sort every $R_n \in \mathcal{R}_N$ in a descending order based on the obtained $U_k^* \big|_{\phi(U_k)=R_n}$ to form F^k.
24: **end for**
25: construct $\mathcal{F}$.

8.2.4 GS Based Second-Stage Task Assignment

When each UAV or subregion has obtained its preference list, the second-stage task assignment is carried out by exploring the GS algorithm [37]. The core parts of the GS algorithm are the propose and reject rules, which are summarized as follows:

Definition 8.3 (Propose Rule) For any subregion $R_n \in \mathcal{R}_N$, it proposes to the top ranking UAV in its preference list F^n, i.e., $\max\{G_n^* \big|_{\phi(R_n)=U_k}, \forall U_k \in F^n\}$.

Definition 8.4 (Reject Rule) For any UAV $U_k \in \mathcal{U}_K$ that has received a matching proposal from a subregion, it can reject the subregion if a better matching candidate exists, and otherwise, the subregion that has not been rejected at the current stage is kept as a matching candidate.

Remark 8.7 (Deferred Acceptance) Assuming that subregion R_n proposes to UAV U_k to form a matching, UAV U_k compares all of the received proposals including the candidate that was kept previously, and will reject subregion R_n if there exists an alternative subregion R_n' which satisfies $R_n' \succeq_{U_k} R_n$. Otherwise, subregion R_n is kept as a matching candidate for UAV U_k, and will be rejected later on if a better candidate appears.

Hence, the matching process is implemented in an iterative fashion, and the detailed process is summarized as follows.

Phase 1: *Matching Preference Initialization*

- Calculate F^n and F^k for each subregion $R_n \in \mathcal{R}_N$ and UAV $U_k \in \mathcal{U}_K$.
- Initialize ϕ as an empty set. Define Φ as the set of subregions that have not been matched, i.e., $\Phi = \mathcal{R}_N$ at the beginning.

Phase 2: *Iterative Matching*

Repeat the following processes iteratively.

- Perform the propose rule for subregions

  - Each subregion $R_n \in \Phi$ proposes to its most preferred UAV in the preference list F^n.

- Perform the reject rule for UAVs

  - Each UAV $U_k \in \mathcal{R}_K$ rejects the subregion which has proposed to it if a better matching candidate exists. Otherwise, the subregion is kept as a matching candidate.

- Update Φ by removing the matched subregions from Φ, and adding the rejected subregions into Φ. Remove any UAV that issues a rejection from the preference list of the rejected subregion.

Until Every subregion $R_n \in \Phi$ has already been accepted by a UAV, or has been rejected by all the UAVs in its preference list.

Phase 3: *Sensing Task Implementation*

Each UAV starts to perform the sensing task based on the matching result obtained in **Phase 2**. Assuming $\phi(R_n) = U_k$, UAV U_k will sense the target points of subregion R_n according to the optimal segment selection and route control strategies derived in Sect. 8.2.2. When all of the subregions have been sensed and the data have been uploaded by UAVs, the MCS carrier can issue a new sensing request, and the task assignment and route planning process will return to **Phase 1**.

Remark 8.8 It is noted that any UAV $U_k \in \mathcal{U}_K$, which cannot satisfy the constraint of sensing latency T_n^{max}, will be directly removed from F^n despite of the preference. On the other hand, if the energy required by UAV U_k to sense subregion $R_n \in \mathcal{R}_N$ exceeds the battery capacity E_k^{max}, subregion R_n will also be directly removed from F^k despite of the preference.

If the K UAVs are completely identical (including the battery level), then the preference of UAV U_k for sensing subregion R_n only depends on the velocity control strategy $\mathcal{V}_{k,n}$, the segment selection strategy $\mathcal{S}_{k,n}$, and the price coefficient $\sigma_{k,n}$. In this case, the route planning problem still has to be solved for each subregion R_n in order to get the optimal velocity control strategy and the segment selection strategy. However, the total complexity will be reduced by K times. Furthermore, if the price coefficient $\sigma_{k,n}$ is the same for any UAV, i.e., $\sigma_{k,n} = \sigma_{k',n}, \forall k \neq k'$, then the second-stage task assignment optimization is neither required.

8.3 Performance Results and Discussions

In this section, we evaluate the joint task assignment and route planning optimization algorithm through numerical results. Simulation parameters are summarized in Table 8.1.

Figure 8.3 demonstrates the relationship among segment energy consumption (e.g., $E^p_{k,l^n_{i,i'}}$), velocity (e.g., $v^{k,n}_{i,i'}$), and segment length (e.g., $d^n_{i,i'}$). It is observed that the segment energy consumption increases linearly with segment length, which is consistent with (8.6). On the other hand, the segment energy consumption firstly decreases as velocity $v^{k,n}_{i,i'}$ increases from $v^{k,n}_{i,i',min}$, because the energy saved per unit velocity increment is higher than the increased energy consumption. The minimum segment energy consumption is achieved when $v^{k,n}_{i,i'}$ reaches the optimal value, i.e., $v^{k,n*}_{i,i'} = 6.68$ m/s. As $v^{k,n}_{i,i'}$ increases beyond $v^{k,n*}_{i,i'}$, the segment energy consumption increases monotonically with velocity. The reason is that the energy consumption increased due to velocity increment begins to dominate the saved energy.

Figure 8.4 compares the DP and GA based route planning algorithms in terms of energy consumption and computation complexity. Simulation results demonstrate that there exists a tradeoff between energy saving gain and complexity reduction. Although the DP-based scheme can achieve a lower energy consumption than the GA-based scheme, its computation complexity increases exponentially with the number of target points increase. If certain performance loss is tolerable, the GA-

Table 8.1 Simulation parameters

Simulation parameters	Value
Segment length $d^n_{i,i'}$	$100 \sim 1000$ m
Propulsion power related coefficient $c_{k,1}$	$0.01 \sim 0.1$
Propulsion power related coefficient $c_{k,2}$	$20 \sim 200$
Price coefficient α_n	$1 \sim 2$
Price coefficient β_n	$0 \sim 1$
Price coefficient $\sigma_{k,n}$	$10^{-4} \sim 10^{-3}$
Accuracy of collected data δ_k	$0 \sim 1$
The amount of data in each subregion D_n	8 Mb ~ 120 Mb
UAV battery capacity E^{max}_k	2×10^4 mAh
Sensing latency constraint T^{max}_n	$400 \sim 1300$ s
Maximum flight velocity $v^{k,n}_{i,i',max}$	20 m/s
Minimum flight velocity $v^{k,n}_{i,i',min}$	2 m/s

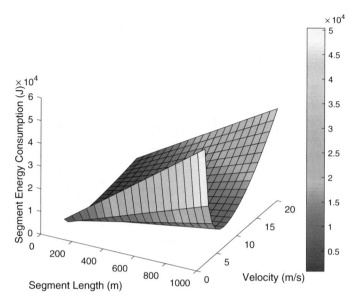

Fig. 8.3 The relationship among segment energy consumption (e.g., $E^p_{k,l^n_{i,i'}}$), velocity (e.g., $v^{k,n}_{i,i'}$), and segment length (e.g., $d^n_{i,i'}$). ($K = 1$, $N = 1$, $c_{k,1} = 0.1$, $c_{k,2} = 200$)

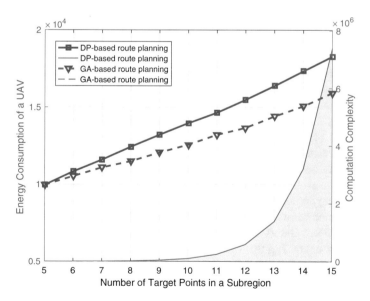

Fig. 8.4 The comparison of DP and GA based route planning algorithms in terms of energy consumption and computation complexity. ($K = 1$, $N = 1$, $c_{k,1} = 0.05$, $c_{k,2} = 150$, $d^n_{i,i'} = 100\,\text{m} \sim 1000\,\text{m}$)

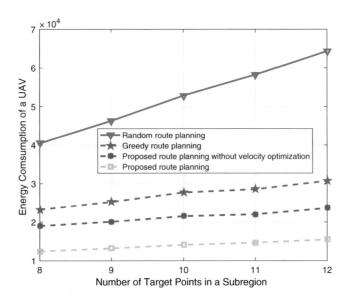

Fig. 8.5 The energy consumption of UAV U_k in subregion R_n versus different numbers of target points, i.e., I_n. ($K = 1$, $N = 1$, $c_{k,1} = 0.05$, $c_{k,2} = 150$, $d_{i,i'}^n = 100\,\text{m} \sim 1000\,\text{m}$)

based scheme is more suitable to handle the large-scale route planning problems with hundreds of target points.

Figure 8.5 shows the energy consumption of UAV U_k in subregion R_n versus different numbers of target points, i.e., I_n. The DP-based route planning algorithm with velocity control optimization is compared to several heuristic algorithms including the DP-based route planning algorithm without velocity control optimization [102], the greedy route planning algorithm without velocity control optimization (the segment with minimum energy consumption is always selected) [103], and the random route planning algorithm without velocity control optimization (the next segment is selected randomly). It is observed that the energy consumption of all the algorithms increases with the number of target points since more energy is required to sense a larger geographic area. Nevertheless, the algorithm achieves the minimum energy consumption since the segment selection and velocity control are jointly optimized from an energy efficiency perspective. The random route planning algorithm achieves the worst performance since energy consumption has not been taken into account during the decision process.

Figure 8.6 shows the impact of sensing latency constraint on the profit of the MCS carrier. For the sake of illustrative purpose, the sensing latency constraint of each subregion is assumed as the same. The DP-based route planning algorithm is compared with greedy route planning and random route planning, where velocity optimization has been considered for all the algorithms. When the sensing latency constraint is in the lower regime, the profit of the MCS carrier is constrained by the number of eligible UAVs which can satisfy the stringent latency requirement. As we

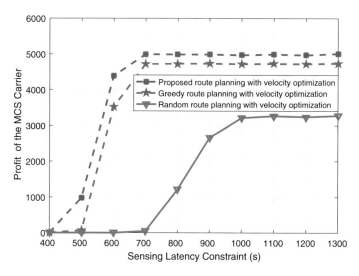

Fig. 8.6 The impact of sensing latency constraint on the profit of the MCS carrier. ($K = 10, N = 5, I_n = 8 \sim 12, \alpha_n = 1 \sim 2, \beta_n = 0 \sim 1, \sigma_{k,n} = 10^{-4} \sim 10^{-3}, D_n = 8\,\text{Mb} \sim 120\,\text{Mb}, c_{k,1} = 0.01 \sim 0.1, c_{k,2} = 20 \sim 200$)

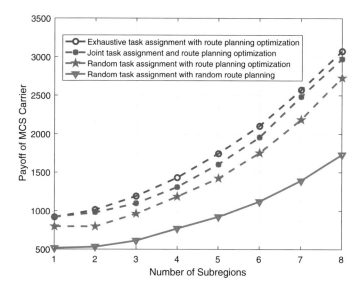

Fig. 8.7 The profit of the MCS carrier versus the number of subregions. ($K = 6, I_n = 8 \sim 15, \alpha_n = 1 \sim 2, \beta_n = 0 \sim 1, \sigma_{k,n} = 10^{-4} \sim 10^{-3}, D_n = 8\,\text{Mb} \sim 120\,\text{Mb}, c_{k,1} = 0.01 \sim 0.1, c_{k,2} = 20 \sim 200$)

loose the sensing latency constraint, the probability of being matched with a more preferred candidate increases significantly with the number of eligible matching candidates increases. This is also called the *diversity gain*, which improves the performances of all the algorithms. Particularly, the algorithm achieves the best performance. The reason behind is that under the same optimal velocity strategy, the DP-based route planning algorithm always achieves the minimum flying distance, which corresponds to the minimum sensing latency, and thus, the maximum profit of the MCS carrier. Furthermore, when the sensing latency constraint is in the higher regime, a performance floor appears since the number of UAVs that are allowed to perform sensing tasks gets saturated, and the corresponding diversity gain varnishes.

Figure 8.7 shows the profit of the MCS carrier versus the number of subregions. The joint task assignment and route planning algorithm is compared with several algorithms including the exhaustive task assignment with route planning optimization, in which the optimal exhaustive matching based on the Monte Carlo approach is utilized to solve the task assignment problem [104], the random task assignment algorithm with route planning, in which the subregions and UAVs are matched randomly, and the random task assignment algorithm without route planning. Velocity optimization has been considered for all the algorithms. With the same route planning optimization, the GS-based task assignment algorithm can achieve up to 96.58% of the optimal performance obtained by the exhaustive task assignment algorithm. Furthermore, Fig. 8.7 shows that the route planning optimization plays an important role in performance improvement. For instance, when $N = 5$, by incorporating route planning, the performance of random task assignment could be improved by 54.24%. The reason is similar to what has been explained in Fig. 8.6. That is, for the same UAV, the DP-based route planning algorithm always achieves the minimum sensing latency, and thus, the maximum profit of the MCS carrier. It is noted that although the random task assignment with route planning optimization achieves a relatively good performance, it is not feasible in practice since the random task assignment decision is not stable. Either the subregions or the UAVs have an incentive to violate the task assignment decision in order to achieve a higher profit.

In this chapter, we investigated the joint task assignment and route planning problem in UAV-aided MCS systems from an energy efficiency perspective. We transformed the original NP-hard joint optimization problem into a two-sided two-stage matching problem, which consists of UAVs on one side, and target subregions on the other side. Then, we employed the DP algorithm to solve the first-stage route planning problem for the construction of the preference list, and employed the GS-based iterative matching algorithm to solve the second-stage task assignment problem. Simulations have been conducted to validate the algorithm under numerous scenarios. We found that the algorithm can achieve superior performance on energy consumption, total profit, and matching satisfaction.

Chapter 9
Energy-Efficient and Secure Resource Allocation for Multiple-Antenna NOMA with Wireless Power Transfer

9.1 Framework of Energy-Efficient and Secure Resource Allocation for Multiple-Antenna NOMA with Wireless Power Transfer

In this section, we firstly provide a detailed description of the system model of multiple-antenna NOMA with wireless power transfer, analytical models and then present the formulation of the energy-efficient and secure resource allocation problem (Fig. 9.1).

9.1.1 System Model

(1) System Assumption

We consider a wireless network consisting of one energy transmitter (EnT) with $N_T > 1$ antennas, K energy harvesting receivers (EHRs) with $N_R > 1$ antennas and an eavesdropper with $N_E \geq 1$ antennas. The EnT is able to charge the EHRs via WPT. After charging, the EHR can utilize the received energy and then transmit the information data to the EnT. Correspondingly, in the WIT phrase, the EHRs become information transmitter (InTs) and the EnT becomes the information receiver (InR). In the following, InT and EHR, EnT and InR are used interchangeably. For security concerns, we assume $N_R > N_E$. It should be noticed that the eavesdropping capability increases with the number of antennas N_E of the eavesdropper. In practice, the system may not know the number of N_E. Thus, the InT may assume N_E as $N_E = N_R - 1$ to ensure security by considering the worst-case scenario. If the assumption is violated, then the eavesdropper is able to eliminate the artificial noise, resulting in throughput of eavesdropper goes to infinite. Therefore, in this work, we assume $N_E = N_R - 1$ and provide corresponding analysis. Meanwhile, the eavesdropper can passively overhearing the information data. The overall bandwidth

© Springer Nature Switzerland AG 2021
Z. Zhou et al., *Green Internet of Things (IoT): Energy Efficiency Perspective*,
Wireless Networks, https://doi.org/10.1007/978-3-030-64054-5_9

Fig. 9.1 System model

Fig. 9.2 Wireless power and information transfer

B is divided into N subchannels, each with bandwidth $W = B/N$. The set of subchannels is denoted as $\mathcal{N}$ and the set of users that uses subchannel n is denoted as $\mathcal{U}_n$. The considered scenario can be applied to the traditional cellular networks or wireless sensor networks. We assume that the CSI between the EnT/InR and EHRs/InTs are perfectly known, but the one of the eavesdropper is unknown. With SIC, some of the co-channel interference will be treated as decodable signals instead of as additive noise.

(2) Wireless Power Transfer
The transmission process of the considered system is depicted in Fig. 9.2. We consider a quasi-static block fading channel model where the channel between the transmitter and receiver is constant for a given transmission block T, and it can vary independently from one block to another. We assume that the whole transmission process including WPT and WIT phrases. As shown in Fig. 9.2, in the first time slot τT, the EnT charges EHR k via WPT and the EHR stores the harvested energy in a rechargeable battery. Then, in the time duration $(1 - \tau)T$, EHR k becomes the InT and sends its own data to the EnT/InR.

Considering the devices are equipped with wireless energy harvesting capability, the energy harvested by EHR k on subchannel n can be considered as follows [105],

$$E_{k,n} = \vartheta \tau T P_n |\mathbf{B}_{k,n}^H \mathbf{H}_{k,n}|^2 = \vartheta \tau T P_n \|\mathbf{H}_{k,n}\|^2 \tag{9.1}$$

where τ is the time fraction of WPT. P_n is the transmit power of the EnT on subchannel n. $\mathbf{H}_{k,n}$ is the channel coefficient matrix from the EnT to the EHR k on subchannel n. $0 < \vartheta < 1$ is the receiver efficiency of WPT, which depends on the hardware features of the receiver. In order to maximize the harvested energy, we design the energy beamforming policy as $\mathbf{B}_k = \frac{\mathbf{H}_{k,n}}{\|\mathbf{H}_{k,n}\|}$, which is named as maximum ratio transmission (MRT). In this work, we do not consider circuit activation energy for the energy harvesting circuit in the EHR. The threshold for activating the energy harvesting circuits is usually considered as a fixed value and such assumption has no impact on addressing the considered problem in this work.

(3) Transmission Protocol
During the data transmission, the received signals at InR and eavesdropper are, respectively, given by,

$$\mathbf{y}_{k,n} = \mathbf{H}_{k,n}\mathbf{x}_{k,n} + \sum_{u\in k}^{U_n} \mathbf{H}_{u,n}\mathbf{x}_{u,n} + \mathbf{n}_{k,n},$$

$$\mathbf{y}_{k,n,E} = \mathbf{G}_{k,n,E}\mathbf{x}_{k,n} + \sum_{u\in k}^{U_n} \mathbf{G}_{u,n,E}\mathbf{x}_{u,n} + \mathbf{n}_{k,n,E}, \tag{9.2}$$

where $\mathbf{H}_{k,n} \in \mathbb{C}^{N_T \times N_R}$ and $\mathbf{G}_{k,n,E} \in \mathbb{C}^{N_E \times N_R}$ are the channel coefficient matrices including the path loss effect between the InT k and InR, and between the InT k and eavesdropper, respectively. $\mathbf{x}_{k,n} \in \mathbb{C}^{N_R \times 1}$ denotes the transmitted signal of InT. $\mathbf{n}_{k,n} \in \mathbb{C}^{N_T \times 1}$ and $\mathbf{n}_{k,n,E} \in \mathbb{C}^{N_E \times 1}$ are the additive white Gaussian noise (AWGN) at InR and the eavesdropper, respectively. The noises follow the distribution $\mathcal{CN}(0, \sigma^2 \mathbf{I}_{N_T})$ and $\mathcal{CN}(0, \sigma^2 \mathbf{I}_{N_E})$, respectively. For the sake of simplicity, a normalized noise variance is assumed for all receivers in the following. To prevent the eavesdropper overhearing the information data, the InT can add artificial noise to the transmission signal in the following way:

$$\mathbf{x}_{k,n} = \mathbf{b}_{k,n}u_{k,n} + \mathbf{V}_{k,n}\mathbf{v}_{k,n}, \tag{9.3}$$

where $u_{k,n}$ is the information bearing signal. Precoding is adopted to improve the system throughput. $\mathbf{b}_{k,n} \in \mathbb{C}^{N_T \times 1}$ is the precoding vector. $\mathbf{v}_{k,n} \in \mathbb{C}^{(N_R-1)\times 1}$ is artificial noise vector whose elements are independent and identically distributed (i.i.d.) complex Gaussian random variables with variance $\sigma^2_{k,n,v}$. Without loss of generality, we define the orthogonal basis $\mathbf{V}_{k,n} \in \mathbb{C}^{N_R \times (N_R-1)}$ for the null space of $\mathbf{H}_{k,n}$ such that $\mathbf{H}_{k,n}\mathbf{V}_{k,n}\mathbf{v}_{k,n} = 0$ and $\mathbf{V}_{k,n}^{\dagger}\mathbf{V}_{k,n} = \mathbf{I}_{N_R-1}$ where $\mathbf{I}_{N_R-1}$ is a $(N_R - 1) \times (N_R - 1)$ identity matrix, i.e., artificial noise will do nothing for the desired receiver. We denote $p_{k,n}$ as the transmit power of InT k on subchannel n, where we have $p_{k,n} = \frac{E_{k,n}}{(1-\tau)T}$, and $\beta_{k,n}$ as the fraction of transmit power. Then, choosing $\mathbf{b}_{k,n} = \beta_{k,n}p_{k,n}\mathbf{H}_{k,n}^{\dagger}/\|\mathbf{H}_{k,n}\|$, such that $u_{k,n}$ lies in the range space of $\mathbf{H}_{k,n}$. As can be seen, the transmitted signal consists of two parts. One is the information bearing signal and the other one the artificial noise. Correspondingly, One important design parameter is the ratio of power allocated to the information bearing signal and the artificial noise. The power of artificial noise vectors can be given by Zhou and McKay [106],

$$\sigma_{k,n,v}^2 = \frac{(1 - \beta_{k,n}) p_{k,n}}{N_R - 1}. \tag{9.4}$$

(4) Secure Capacity

Let us first consider a fixed decoding order of the InTs' messages at the InR, according to their index k. The InR has capabilities of multi-user detection (MUD) and SIC.

We adopt the descending order of channel gains as the decoding order, in order to improve the throughput of weak-channel users and enhance user fairness [107]. We sort all U_n InTs on each subchannel n in descending order of channel gains, $|\mathbf{H}_{1,n}| \geq |\mathbf{H}_{2,n}| \geq \ldots \geq |\mathbf{H}_{U_n,n}|$. The 1st signal (the strongest user) is detected and decoded first by treating all the other users' signals as noise. For the U_nth user (the weakest user), the BS successively decodes and removes all the interference of strong users from 1 to $U_n - 1$, before decoding the signals of the weakest user. In general for the kth user, the BS can remove the received signals from the 1st to the $k - 1$th users, and consider the $k + 1$th to the U_nth users as noise.

Correspondingly, the achievable uplink data rate of k on subchannel n, $C_{k,n}$, can be expressed as

$$C_{k,n} = W T (1 - \tau) \log_2 (1 + \gamma_{k,n}). \tag{9.5}$$

where $\gamma_{k,n}$ is the uplink SINR of k . It can be given as

$$\gamma_{k,n} = \frac{\beta_{k,n} p_{k,n} \|\mathbf{H}_{k,n}\|^2}{\sigma^2 + \sum_{u=k+1}^{U_n} \beta_{u,n} p_{u,n} \|\mathbf{H}_{u,n}\|^2}, \tag{9.6}$$

where σ^2 is the noise variance. $p_{k,n}$ is the transmit power of k on subchannel n. In this work, all the harvesting energy can be used for transmitting data.

In practice, it may not be easy to know the capabilities of the eavesdropper on overhearing, therefore, here we also consider the case that all the information received at eavesdropper are useful for overhearing. Moreover, in many practical cases, the eavesdropper is passive and the CSI between the InTs and eavesdropper is unknown. Therefore, the data rate of the eavesdropper of $C_{k,n,E}$ should be considered a random variable, which can be given as

$$C_{k,n,E} = W T (1 - \tau) \log_2 \left(1 + \Gamma_{k,n,E}\right). \tag{9.7}$$

where $\Gamma_{k,n,E}$ is the SINR. We also assume that the eavesdropper is much closer to the InT than the desired InR, the eavesdropper noise is negligible. Based on (9.2), (9.3) and (9.4) it can be given as

$$\Gamma_{k,n,E} = \frac{N_R - 1}{1 - \beta_{k,n}} \tilde{\mathbf{g}}^\dagger \left(\tilde{\mathbf{G}} \tilde{\mathbf{G}}^\dagger\right)^{-1} \tilde{\mathbf{g}}, \tag{9.8}$$

where $\tilde{\mathbf{g}} = \mathbf{G}_{k,n,E}\mathbf{b}_{k,n}$ and $\widetilde{\mathbf{G}} = \mathbf{G}_{k,n,E}\mathbf{V}_{k,n}$. $\mathbf{G}_{k,n,E}$ is the channel co-efficient between k and eavesdropper on subchannel n. Note the eavesdropper's capability for decoding the message is overestimated here. A worst-case assumption is made for obtaining the data rate of eavesdropper from the legitimate users' perspective. That is, the eavesdropper consider every message is useful and is able to obtain related information of the transmitter before it attempts to decode the message. As a matter of fact, the eavesdropper may or may not know the users' decoding order and the resource allocation policy, and may or may not know the transmitter before it attempts to decode the message. However, the InTs do not have the knowledge about the eavesdropper, since the eavesdropper would not inform its ability and CSI. Thus, the worst-case assumption is adopted here for estimating the data rate of eavesdropper due to the conservativeness mandated by the security studies. Correspondingly, the secrecy capacity of user k on subchannel n can be expressed as

$$C_{k,n}^s = (C_{k,n} - C_{k,n,E})\phi(C_{k,n} > C_{k,n,E}), \tag{9.9}$$

where

$$\phi(C_{k,n} > C_{k,n,E}) = \begin{cases} 1, \text{ if } C_{k,n} > C_{k,n,E}; \\ 0, \text{ otherwise.} \end{cases} \tag{9.10}$$

It is worth mentioning that when the CSI of the eavesdropper is known, it is possible to set the target secrecy data rate and control the channel capacity to via power or subchannel allocation to match the channel conditions. However, as $C_{k,n,E}$ should be considered a random variable in this context and we cannot obtain $C_{k,n}^s$ perfectly. Correspondingly, a secrecy outage occurs when a defined target secrecy data rate $R_{k,n}$ exceeds $C_{k,n}^s$, i.e., the message can be delivered securely and successfully when $R_{k,n} < C_{k,n}^s$. Therefore, instead of ergodic capacity, in the following we study the performance w.r.t. the target secrecy data rate $R_{k,n}$ instead of $C_{k,n}^s$, which will be elaborate in the following section.

(5) Energy Consumption Model
In this work, we consider the maximum efficiency of energy utilization, that is, all the harvested energy is used for data transmission such that no energy is wasted. We denote $\mathcal{P}(\mathbf{P}, \tau)$ as the total energy consumption in a time block T and it can be expressed as

$$\mathcal{P}(\mathbf{P}, \tau) = \sum_{n=1}^{N} \tau \nu P_n + P_c T, \tag{9.11}$$

where $\mathbf{P}$ is a collection of all power elements, and P_c is the static power consumption, such as the power consumption on baseband and RF chain for antenna [109]. ν is the factor standing for the nonlinear power amplifier effect.

9.1.2   Problem Formulation

Based on our system model of secrecy data rate and energy consumption, we are able to investigate the secure and energy-efficient resource allocation problem for the considered system. In this Section, we present the formulation of EE problem and also explain the practical constraints. First, to facilitate the presentation of EE formulation, we define a subchannel allocation indicator $\omega_{k,n}$ as follow,

$$\omega_{k,n} = \begin{cases} 1, & \text{if subchannel } n \text{ is allocated to the user } k; \\ 0, & \text{otherwise.} \end{cases} \tag{9.12}$$

Next, to formulate the optimization problem, we utilize the definition the EE of the considered system in bits/J as follows [111]:

$$\Sigma(\mathbf{P}, \tau, \boldsymbol{\omega}) = \frac{\mathcal{U}(\mathbf{P}, \tau, \boldsymbol{\omega})}{\mathcal{P}(\mathbf{P}, \tau, \boldsymbol{\omega})}, \tag{9.13}$$

where $\mathcal{P}(\mathbf{P}, \tau, \boldsymbol{\omega})$ is interchangeable with $\mathcal{P}(\mathbf{P}, \tau)$, and

$$\mathcal{U}(\mathbf{P}, \tau, \boldsymbol{\omega}) = \sum_{k=1}^{K} \sum_{n=1}^{N} \omega_{k,n} R_{k,n} \mathbf{Pr}\{R_{k,n} < C_{k,n} - C_{k,n,E} | \Delta\}, \tag{9.14}$$

where Δ is the CSI of user k on subchannel n. To this end, we jointly optimize duration τ, allocation indicators $\boldsymbol{\omega}$, and power allocation $\mathbf{P} = \{P_1, \ldots, P_n, \ldots, P_N\}$. The optimization problem can be formulated as follows:

$$\mathbf{P1}: \max_{\mathbf{P}, \tau, \boldsymbol{\omega}} \Sigma(\mathbf{P}, \tau, \boldsymbol{\omega}), \tag{9.15}$$

s.t.

$$\mathbf{C1}: \mathbf{Pr}\{R_{k,n} \geqslant C_{k,n} - C_{k,n,E} | \Delta\} \leqslant \varepsilon, \forall k \in \mathcal{K}, \forall n \in \mathcal{N}$$

$$\mathbf{C2}: \sum_{n=1}^{N} P_n \leqslant P_{b,max},$$

$$\mathbf{C3}: \sum_{n=1}^{N} \frac{\vartheta \tau P_n \|\mathbf{H}_{k,n}\|^2}{(1-\tau)} \leq P_{u,max}, \forall k \in \mathcal{K}$$

$$\mathbf{C4}: \sum_{n=1}^{N} R_{k,n} \geqslant R_{min}, \forall k \in \mathcal{K}$$

$$\mathbf{C5}: \sum_{k=1}^{K} \omega_{k,n} \leq L, \forall n \in \mathcal{N}$$

$$\mathbf{C6}: \omega_{k,n} \in \{0, 1\}, \text{ and } P_n \geq 0,$$

$$\mathbf{C7}: 0 < \tau \leq 1, \tag{9.16}$$

Our goal is to maximize the system EE under a set of practical constraints. **C1** is the QoS metric for communication security, where ε denotes the maximum tolerable secrecy outage probability. **C2** and **C3** impose limitations on the power consumption and ensure the feasibility of power allocation solutions. Specifically in **C2**, total transmit power for WPT at EnT is no larger than a maximum power limit $P_{b,max}$. In **C3**, we use $\frac{\vartheta \tau P_n \|\mathbf{H}_{k,n}\|^2}{(1-\tau)}$ to represent the power value of UE k on subchannel n, i.e., $p_{k,n}$. Each user's transmit power for uplink data transmission cannot exceed its maximum power limit $P_{u,max}$. In **C4**, R_{min} is the minimum system secrecy rate requirement, and a balance between EE and secrecy outage capacity can be achieved by varying R_{min}. Constraint **C5** indicates that the maximum number of allocated users on each subchannel is up to L. Note that parameter L is an integer value no less than $\frac{K}{N}$, otherwise the problem is infeasible. Constraints **C6** to **C7** are the boundary for optimization variables. We remark that in **P1** we do not explicitly impose an extra constraint to indicate the connection between variables P_n and $\omega_{k,n}$. It reads, for any subchannel n, if $\sum_{k=1}^{K} \omega_{k,n} > 0$ then $P_n > 0$, and $\sum_{k=1}^{K} \omega_{k,n} = 0$ indicates $P_n = 0$. Note that both of the unwanted decisions, $\sum_{k=1}^{K} \omega_{k,n} > 0$ with $P_n = 0$ and $\sum_{k=1}^{K} \omega_{k,n} = 0$ with $P_n > 0$, will not be presented in the optimum. This is because, either the case of that InTs have been allocated to subchannels but with zero assigned power and zero rate, or power has been consumed but no rate value contributes to the objective, is clearly not optimal.

We remark that the optimization of user grouping is an important aspect in NOMA resource allocation. In this work, determining which users to be assigned to which subcarrier is an outcome of optimization, i.e., decided by the binary variables $\omega_{k,n}$. This is different from some previous works in the literature, in which the co-channel allocated users are predefined in prior of optimization. In the above optimization of problem, the constraints in secrecy outage probability **C1**, users' power constraints **C3**, minimum-rate constraints **C4**, and constraints **C5** together with the objective function provide major effect in determining user-subcarrier allocation.

It can be noticed that the formulated problem is with a non-convex structure. The objective function in a fractional program is a ratio of two functions of the optimization variables. In order to make the problem solvable, we transform the objective function and approximate the transformed objective function. In the following, we first transform the optimization problem by utilizing the idea from fractional programming and then address the transformed problem.

9.2   The Energy-Efficient and Secure Resource Allocation Scheme

In this section, we introduce the energy-efficient and secure resource allocation scheme. First, we transform the optimization problem. Then, we propose an algorithmic solution to solve the nonconvex optimization problem.

9.2.1   Transformation of the Optimization Problem

In order to make the problem tractable, we utilize the idea from the fractional programming and transform the objective function and approximate the transformed objective function in order to simplify the problem. We can apply the nonlinear fractional programming method to solve the formulated problem [34]. As can be found in the Sect. 9.2.2, it can be found that $\Sigma(\mathbf{P}, \tau, \boldsymbol{\omega})$ can be transformed to a quasi-concave function over the decision variable, then we define the maximum energy efficiency q^* of the considered system and the following theorem can be arrived.

Theorem 9.1 *The maximum EE q^* can be achieved if and only if*

$$\mathbf{U}(\mathbf{P}^*, \tau^*, \boldsymbol{\omega}^*) - q^* \mathcal{P}(\mathbf{P}^*, \tau^*, \boldsymbol{\omega}^*) = 0, \tag{9.17}$$

The proof is similar to the one in [111]. To find the optimal q^*, the iterative algorithm with guaranteed convergence in [34] can be applied. To make it adapted for solving our problem, we use the framework, and adjust the procedure, as shown in Algorithm 9.1. The proof of convergence of Algorithm 9.1 can be found in [34]. During each iteration, we need to solve the following problem for a given q,

---

Algorithm 9.1 Iterative algorithm for obtaining q^*

1: Set maximum tolerance δ;
2: **while** (not convergence) **do**
3:    Solve the problem (9.18) for a given q and obtain subchannel, power and time allocation $\{\mathbf{P}', \tau', \boldsymbol{\omega}'\}$;
4:    **if** $\mathcal{U}(\mathbf{P}', \tau', \boldsymbol{\omega}') - q\mathcal{P}(\mathbf{P}', \tau', \boldsymbol{\omega}') \leq \delta$ **then**
5:       Convergence = true;
6:       **return** $\{\mathbf{P}^*, \tau^*, \boldsymbol{\omega}^*\} = \{\mathbf{P}', \tau', \boldsymbol{\omega}'\}$ and obtain q^* by (9.17);
7:    **else**
8:       Convergence = false;
9:       **return** Obtain $q = \frac{\mathcal{U}(\mathbf{P}', \tau', \boldsymbol{\omega}')}{\mathcal{P}(\mathbf{P}', \tau', \boldsymbol{\omega}')}$;
10:   **end if**
11: **end while**

$$\mathbf{P2}: \quad \max_{\mathbf{P}, \tau, \boldsymbol{\omega}} \mathbf{U}(\mathbf{P}, \tau, \boldsymbol{\omega}) - q\mathcal{P}(\mathbf{P}, \tau, \boldsymbol{\omega}),$$

$$\text{s.t. } \mathbf{C1} - \mathbf{C7}. \tag{9.18}$$

We are able to replace the "$\leq$" in **C1** by a "=" sign and substitute it into the optimization problem, and the resulting optimization problem can be viewed as a restricted version of the original problem since it has a smaller feasible set [113].

In order to address the formulated problem, next, we aim at obtaining the secrecy capacity $R_{k,n}$. Correspondingly, the following conclusion can be achieved.

Theorem 9.2 *Assuming the channel between the InT and eavesdropper is Rayleigh fading, the equivalent secure data rate for InT k is given by*

$$R_{k,n} = WT(1 - \tau)\left[\log_2(1 + \gamma_{k,n}) - \log_2\left(\frac{\beta_{k,n}^* \Omega_E^{1/2}}{1 - \beta_{k,n}^*} \right) \right]^+,$$

(9.19)

where

$$\Omega_E = (N_R - 1)F_{z_h}^{-1}(\varepsilon),$$

$$\beta_{k,n}^* = \frac{1}{\sqrt{2\Omega_E}}.$$

(9.20)

We also assume that $\gamma \gg 1$. $F_{z_h}^{-1}(\varepsilon)$ *is the inverse function of* $F_{z_h}(z) = \varepsilon$, *and* ε *denotes the maximum tolerable secrecy outage in* **C1**. $F_{z_h}(z)$ *is given by*

$$F_{z_h}(z) = \frac{\sum_{n=0}^{N_E-1} \binom{N_R-1}{n} 2z^n}{(1 + z)^{N_R-1}}$$

$$- \frac{\sum_{n=0}^{N_E-1} \sum_{m=0}^{N_E-1} \binom{N_R-1}{n}\binom{N_T-1}{m} z^{m+n}}{(1 + z)^{2N_R-2}}.$$

(9.21)

Proof The detailed proof is given in [114].   □

From **Theorem** 9.2, it can be found that the SINR from the InT to the eavesdropper becomes a constant value at high SNR and it is independent of the decision variables, which simplifies the derivation of the resource allocation scheme. With the above analysis, in the following, we can address the transformed optimization problems.

9.2.2   Proposed Algorithmic Solution

The transformed problem **P2** is still with a non-convex structure. Tackling the mixed non-convex and combinatorial optimization problem requires a prohibitively high complexity. Although a possible solution can be obtained when addressing such problem in the dual domain. However, for the formulated optimization problem, it may result in a large duality gap between the primal and the dual problem as the non-linear mixed integer programming is involved. In addition, directly solving the whole problem **P2** by developing simple heuristics, e.g., greedy algorithms, may also lead to a large optimality loss in the optimization process. In this chapter, we investigate the problem's insights, and derive analytical results for solving **P2**. We firstly relax the problem and decompose the whole optimization process, then we design an iterative search method with guaranteed convergence, and solve the problem to the optimum at each iteration. The developed algorithm reuses the optimal solution of each iteration as much as possible. Next, we present the algorithm design in details.

The proposed algorithmic solution is based on the following analytical results. Firstly, one can observe that from **Theorem** 9.2, the entity $\log_2 \left(\frac{\beta_{k,n}^* \Omega_E^{1/2}}{1 - \beta_{k,n}^*} \right)$ in $R_{k,n}$ is approximately a constant. Constraint **C1** can be absorbed into objective function in **P2** or **P1**, without loss optimality. Hence, we can rewritten $R_{k,n}$ as $WT(1 - \tau) \left[\log_2(1 + \gamma_{k,n}) - \tilde{V}_{k,n} \right]^+$, where $\tilde{V}_{k,n} = \log_2 \left(\frac{\beta_{k,n}^* \Omega_E^{1/2}}{1 - \beta_{k,n}^*} \right)$ is seen as a parameter. Secondly, if we consider the rate function in uplink NOMA, also shown in other existing works, e.g., [107, 108], the summation of $\log_2(1 + \gamma_{k,n})$ over users on each subchannel can be represented as the following form.

Lemma 9.1 *In uplink NOMA,* $\sum_{k=1}^K \log_2(1 + \gamma_{k,n}) = \log_2 \frac{\sum_{k=1}^K \beta_{k,n} p_{k,n} \|\mathbf{H}_{k,n}\|^2}{\sigma^2}$, $\forall n \in \mathcal{N}$.

Proof For an arbitrary subchannel n, $\sum_{k=1}^K \log_2(1 + \gamma_{k,n}) = \log_2(1 + \frac{\beta_{1,n} p_{1,n} \|\mathbf{H}_{1,n}\|^2}{\sum_{k=2}^K \beta_{k,n} p_{k,n} \|\mathbf{H}_{k,n}\|^2 + \sigma^2}) + \ldots, + \log_2(1 + \frac{\beta_{K,n} p_{K,n} \|\mathbf{H}_{K,n}\|^2}{\sigma^2})$ reads

$$\log_2 \left(\frac{\sum_{k=1}^K \beta_{k,n} p_{k,n} \|\mathbf{H}_{k,n}\|^2 + \sigma^2}{\sum_{k=2}^K \beta_{k,n} p_{k,n} \|\mathbf{H}_{k,n}\|^2 + \sigma^2} \times \frac{\sum_{k=2}^K \beta_{k,n} p_{k,n} \|\mathbf{H}_{k,n}\|^2 + \sigma^2}{\sum_{k=3}^K \beta_{k,n} p_{k,n} \|\mathbf{H}_{k,n}\|^2 + \sigma^2} \times, \right.$$
$$\left. \ldots, \times \frac{\sum_{k=K-1}^K \beta_{k,n} p_{k,n} \|\mathbf{H}_{k,n}\|^2 + \sigma^2}{\beta_{K,n} p_{K,n} \|\mathbf{H}_{K,n}\|^2} \times \frac{\beta_{K,n} p_{K,n} \|\mathbf{H}_{K,n}\|^2}{\sigma^2} \right) \tag{9.22}$$

Eliminating the same entities in the denominator and numerator, (9.22) is equal to $\log_2 \frac{\sum_{k=1}^K \beta_{k,n} p_{k,n} \|\mathbf{H}_{k,n}\|^2}{\sigma^2}$, hence the conclusion. Note that the result holds for arbitrary decoding orders, and it is independent with the optimization outcome of user allocation on subchannel n. □

Based on the above two observations, we can have our next analytical result. For any given τ and q, if **C5** is temporarily removed or relaxed from **P2**, also replacing all entities $\log_2 \left(\frac{\beta_{k,n}^* \Omega_E^{1/2}}{1 - \beta_{k,n}^*} \right)$ and $\log_2(1 + \gamma_{k,n})$ by the derived new forms, we formulate a relaxed version of **P2** in **P3**. We next show **P3** can be solved efficiently.

$$\textbf{P3}: \max_{\textbf{P} \geq 0} WT(1-\tau) \sum_{n=1}^{N} \log_2 \left(\frac{\sum_{k=1}^{K} \alpha_{k,n} P_n}{\sigma^2} \right) - \sum_{k=1}^{K} \sum_{n=1}^{N} \tilde{V}_{k,n}$$

$$- q \left(\sum_{n=1}^{N} \tau v P_n + P_c T \right),$$

$$\text{s.t. } \textbf{C2}: \sum_{n=1}^{N} P_n \leq P_{b,max},$$

$$\textbf{C3}: \sum_{n=1}^{N} \frac{\vartheta \tau P_n \|\mathbf{H}_{k,n}\|^2}{(1-\tau)} \leq P_{u,max}, \ \forall k \in \mathcal{K}$$

$$\textbf{C4}: \sum_{n=1}^{N} WT(1-\tau) \log_2 \left(1 + \frac{\alpha_{k,n} P_n}{\sigma^2 + \sum_{u=k+1}^{K} \alpha_{u,n} P_n} \right)$$

$$- \tilde{V}_{k,n} \geq R_{min}, \ \forall k \in \mathcal{K} \tag{9.23}$$

where $\alpha_{k,n} = \beta_{k,n} \vartheta \tau / (1-\tau) \|\mathbf{H}_{k,n}\|^2 \|\mathbf{H}_{k,n}\|^2$, and $WT(1-\tau) \log_2 (1 + \frac{\alpha_{k,n} P_n}{\sigma^2 + \sum_{u=k+1}^{K} \alpha_{u,n} P_n}) - \tilde{V}_{k,n} = R_{k,n}$ in **C4**. Note that only power is the optimization variable in **P3**, and the binary indicators $\omega_{k,n}$ are no longer needed due to the absence of **C5**. We then show the convexity of **P3** below.

Lemma 9.2 *P3 is convex.*

Proof For the objective function, one can observe its concavity. Constraints **C2** and **C3** are linear. For **C4**, we derive the second derivative for function $f(P_n) = \log_2 (1 + \frac{\alpha_{k,n} P_n}{\sigma^2 + \sum_{u=k+1}^{K} \alpha_{u,n} P_n})$. According to the fact that $f''(P_n) < 0$, we therefore conclude the convexity of **P3**.   □

The global optimum of a convex problem, e.g., **P3**, can be obtained efficiently by either using standard convex optimization tools or by applying KKT conditions. Towards the optimum of **P2**, two aspects can be considered from **P3**. First, given the same q, if the optimal solution in **P3** does not violate **C5**, it is also optimal for **P2**. This is typically observed in many cases that the maximum number of allocated InTs per subchannel, i.e., parameter L, is not too restricted. On the other side, even if **C5** is violated, the similar structure in both problems could lead to high correlation between optimal solutions in **P2** and **P3**, namely, the optimal decisions in **P3** would also be favorable in **P2**. This motivates us to consider that, instead of directly addressing **P1** or **P2**, we can approximate it through **P3** firstly, then derive a near-optimal solution by reusing the optimal solution of **P3** as much as possible.

Based on aforementioned considerations and derived results, we propose an iterative searching scheme in Algorithm 9.2 to solve the original problem **P1**. Algorithm 9.2 is based on the framework of Algorithm 9.1. It updates q iteratively, and for each scanned q, the algorithm finds τ by applying bisection search in Line 3 to 6. In each iteration of bisection search, a convex problem **P3** is efficiently solved in Line 4 when τ and q have been updated. The majority of Algorithm 9.1 is integrated in Line 11 to 18, to decide whether the optimal q is achieved in Line 13, or update suboptimal q in Line 16. According to the proof of [34], and observing the same algorithmic structure between Algorithm 9.1 and 9.2, we can conclude the following observation.

Corollary 9.1 *Convergence is guaranteed in Algorithm 9.2.*

When the major loop terminates at Line 18, if there is no violation for **C5** during iterations, the outcome, $\mathbf{P}^*$, τ^*, ω^*, of the algorithm in Line 20 is an optimal solution for **P1**. Otherwise, the power solution $\mathbf{P}'$ and the corresponding allocation indicator ω' will provide an upper bound for the global optimum of **P1**. This conclusion is formalized blow.

Lemma 9.3 *At the termination of Algorithm 9.2, solution $\{\mathbf{P}', \tau', \omega'\}$ is an upper bound for the optimum of P1.*

Proof Let τ^1, V^1 denote the globally optimal τ and the objective value in **P2**, respectively. Suppose q^1 is optimal, according to Theorem 9.1, q^1 will lead to the equivalence between **P2** and **P1**. If we set τ^1 and q^1 as the input in **P3**, the resulting optimal objective value of **P3**, denoted as V^2, has $V^2 \geq V^1$, because **P3** is a relaxed version of **P2**. In Algorithm 9.2, the finalized τ' and q in Line 22 corresponds to a resulting objective value in **P3**, denoted as V^3. We then conclude $V^3 \geq V^2$. The reason behind is that the optimization procedure for searching τ and q in Algorithm 9.2 is optimal. If τ^1 and q^1 are not obtained by Algorithm 9.2 at the termination, it means τ^1 and q^1 are not the best choice in optimization, and will not lead to higher objective value than obtained τ' and q. Thus, from the results of $V^3 \geq V^1$ and the equivalence between **P1** and **P2** in EE, the finalized solution $\{\mathbf{P}', \tau', \omega'\}$ in Algorithm 9.2 enables an upper bound of **P1**.   □

Finally, in Line 23, $\mathbf{P}'$ and ω' can be adjusted by simple heuristic steps, e.g., the method used in [110], to provide a feasible solution, in general suboptimal, for **P1**. The proposed Algorithm 9.2 has polynomial-time complexity. The majority of the computations are in two nested loops. The outer loop (also refer to as the framework of Algorithm 9.1 in Sec. IV. A) can be proved to have a linear-time complexity [34]. The inner loop of Algorithm 9.2 consists of one-dimension bisection search for τ, associated with solving a convex problem P3 in each iteration. The convex problem P3 can be solved by the barrier-based interior-point method with polynomial-time complexity in the worst case [35]. The complexity of bisection search is much moderate than solving a convex problem. Hence, overall the proposed Algorithm 9.2 is of polynomial-time complexity.

We remark that identifying the solution existence for an optimization problem is an important aspect in algorithm design. If the original problem P1 is infeasible, by solving the relaxed convex problem **P3**, Algorithm 9.2 can efficiently provide a simple feasibility check for the solution existence, which is much more easier than solving a non-convex problem **P1** to answer the feasibility. In general the proposed Algorithm 9.2 is able to provide feasible solutions for solving the formulated problem if the original problem **P1** is feasible, i.e., at least one feasible solution exists. This is because in Algorithm 9.2, instead of relying on heuristic procedures, the majority of the optimization is to iteratively and optimally solve a convex problem **P3** which is a relaxed version of the original non-convex problem **P1**. If the solution exists for **P1**, then solving its relaxed problem **P3** is feasible since the feasible region becomes larger.

Algorithm 9.2 Iterative algorithm for solving **P1**

1: **Initialize**: tolerance δ, $Converge$ = false, $Violate$ = false, and $q = 0$;
2: **while** $Converge$ = false **do**
3:     **Bisection search** for τ **do**
4:       Solve **P3** for current q and τ
5:       Obtain optimal power solution **P** for **P3**
6:     **until** Maximum EE under the current q is achieved at τ' and $\mathbf{P}'$
7:     Convert $\mathbf{P}'$ to its corresponding channel indicators $\boldsymbol{\omega}'$
8:     **if** (**C5** is violated in $\boldsymbol{\omega}'$ and $\mathbf{P}'$) **then**
9:       $Violate$ = true
10:    **end if**
11:    **if** $\mathcal{U}(\mathbf{P}', \tau', \boldsymbol{\omega}') - q\mathcal{P}(\mathbf{P}', \tau', \boldsymbol{\omega}') \leq \delta$ **then**
12:      $Converge$ = true
13:      **return** $\{\mathbf{P}^*, \tau^*, \boldsymbol{\omega}^*\} = \{\mathbf{P}', \tau', \boldsymbol{\omega}'\}$ and update q by (9.17)
14:    **else**
15:      $Converge$ = false
16:      Update $q = \frac{\mathcal{U}(\mathbf{P}',\tau',\boldsymbol{\omega}')}{\mathcal{P}(\mathbf{P}',\tau',\boldsymbol{\omega}')}$
17:    **end if**
18: **end while**
19: **if** $Violate$ = false **then**
20:    **Output 1:** Optimal solution $\{\mathbf{P}^*, \tau^*, \boldsymbol{\omega}^*\}$ for **P1**
21: **else**
22:    **Output 2:** Upper bound $\{q, \mathbf{P}', \tau', \boldsymbol{\omega}'\}$ for **P1** or **P2**
23:    For τ' and q up to date, convert $\mathbf{P}', \boldsymbol{\omega}'$ to a feasible solution $\bar{\mathbf{P}}, \bar{\boldsymbol{\omega}}$ for **P1**
24:    **Output 3:** Suboptimal solution $\bar{\mathbf{P}}, \bar{\boldsymbol{\omega}}$ for **P1**
25: **end if**

9.3 Performance Evaluation

9.3.1 Improve Secure Data Rate

In this section, the performance of the proposed scheme is evaluated and illustrated. In the simulations, we consider one energy transmitter/information receiver and multiple users (EHRs/InTs) and the distance between the EnT and users are about 200 m. The bandwidth is 3 MHz. Some key power consumption parameters related to P_c are mainly from [109] and are also given in Table 9.1. As for wireless power transfer, we assume that the energy conversion efficiency of WPT is $\vartheta = 0.5$. A distance-depended path loss model is considered. To evaluate the performance of proposed scheme (P-NOMA), we have implemented a previous NOMA power and channel allocation scheme called "fractional transmit power control" (FTPC) and an OFDMA scheme with FTPC (OFDMA) [112, 113]. For the implemented OFDMA scheme, each user can only be assigned to one subchannel. In the FTPC, the set of multiplexed users U_n for each n is determined by a greed-based user grouping strategy, where $|U_n| = M$. Based on the user allocation, the FTPC method is then used for power allocation. In considered fractional power control schemes, the users with inferior channel condition will be allocated with more power for the fairness consideration [112]. In addition, we also compare our proposed scheme with equal transmit power allocation scheme (ETPA) and equal time allocation scheme (ETTA). In ETPA, NOMA system is considered and the transmit power on each subchannel is equal while the other schemes are the same as the proposed one. In ETTA, the proposed power allocation and subchannel allocation are used while overall time slot is divided equally for data transmission and power delivery.

In Fig. 9.3, we plot the secure data rate (bps/Hz) versus the transmit power of the EnT. In this figure, we compare the performance of the P-NOMA with that of the OFDMA and ETTA to show the necessity and advantages of our proposed scheme in NOMA and the effectiveness of proposing time allocation scheme. It can be observed that the secure data rate is incremented when the transmit power increases. As the transmit power becomes larger, the secure data rate continues to increment. In NOMA systems, our proposed algorithm performs better than ETTA. Both algorithms in NOMA outperforms the OFDMA system. For example, our proposed resource allocation scheme achieves up to 25% better performance than the ETTA and is about 50% better than that of OFDMA.

Table 9.1 Simulation parameters

Parameter	Value
N	64 for OFDMA case
L	3
$P_{b,max}$	46 dBm
$P_{u,max}$	23 dBm
R_{min}	0.1 bit/s/Hz
η	0.5
ε	10^{-7}
P_{DAC}, P_{ADC}	10 mW
P_{filt}, P_{filr}	2.5 mW
P_{mix}	30.3 mW
P_{syn}	50 mW
P_{LNA}	20 mW
P_{IFA}	3 mW

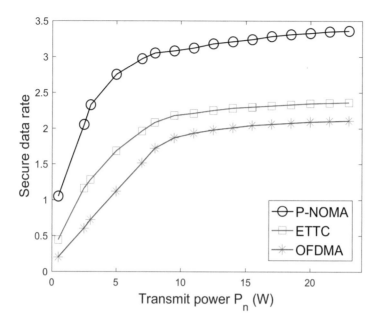

Fig. 9.3 Secure data rate versus transmit power of the EnT

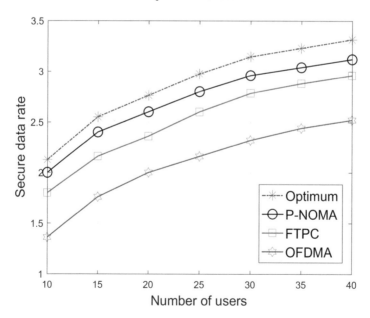

Fig. 9.4 Secure data rate versus number of users/InTs in the system

Figure 9.4 shows the secure data rate (bps/Hz) versus the number of users. In this figure, we also compare the performance of the P-NOMA with that of the FTPC and

OFDMA to show the advantages of our proposed scheme. In addition, we also plot the optimal solution obtained by exhaustive search and named it as "Optimum". It can be observed that the secure data rate increases when the number of users grows. As the number of users becomes larger, the secure data rate continues to increment, while the rate of growth becomes slower, as expected from the way to obtain the secure data rate. From this figure, the performance of our proposed allocation scheme is better comparing with the other two schemes. For example, our proposed resource allocation scheme achieves 10% better performance than the FTPC when the number of users is 25, and is about 50% better than that of OFDMA.

9.3.2 Improve the Energy Efficiency

In Fig. 9.5, the performance of EE (bits/J/Hz) is evaluated with the number of users with the same constraints of Fig. 9.4. To illustrate the advantages of our proposed scheme, we also compare our P-NOMA with the FTPC and OFDMA. It is shown that the EE increases when the number of the users grows. As the number of users grows larger, the EE continues to increase. The trend of the curves is similar to the secure data rate curves in Fig. 9.4 due to the EE formulation in (9.13). In addition, from Fig. 9.5, we can see that the performance of NOMA system with the proposed resource allocation algorithms is better than the OFDMA scheme. For example, when the number of users is 30, the EE of the proposed scheme is about 60%

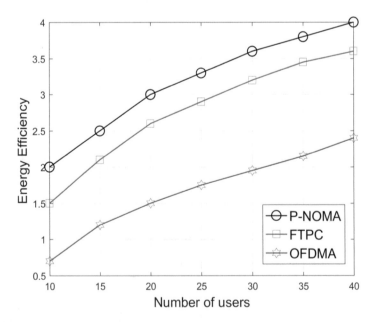

Fig. 9.5 Energy efficiency versus number of users/InTs in the system

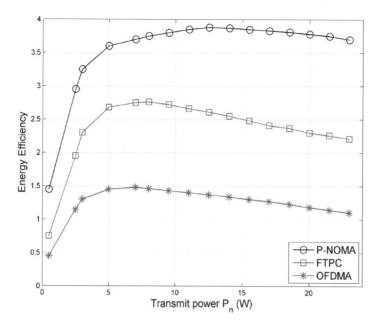

Fig. 9.6 Energy efficiency versus transmit power of the EnT

more than that of OFDMA scheme, and is 15% more than that of FTPC scheme. This is mainly due to the fact that in OFDMA scheme, one subchannel can only be used by one user which decrease the efficient usage of frequency bandwidth. Consequently, spectrum resources cannot be fully utilized. For different subchannel power allocation schemes.

Figure 9.6 plots the EE by changing the transmit power P_n and allocated time slot τ. The performance of the proposed scheme with optimal time allocation (P-NOMA) is compared with the one of OFDMA with optimal time allocation, and the one of the proposed scheme with ETTA, e.g. $\tau = 1/2T$. By the comparing these three curves, we can observe that with the increase of transmit power, the EE of the system first ascends and then descends. Similar to the results in other figures, Fig. 9.6 shows that the transmit power has an optimal value, which confirms the advantages and necessity of power allocation scheme. Moreover, it can be seen that our proposed time allocation scheme can obtain additional EE gain when comparing with the equal time allocation scheme. Last but not the least, we can find that the EE of our proposed scheme is the highest among all three, which shows the advantages of our proposed algorithms over the traditional schemes.

In Fig. 9.7, we present the total EE in bit/J/Hz versus the time allocation parameter τ. In this figure, we vary the value of first time slot τ and present the EE performance of different resource allocation schemes. Moreover, we also consider the whole time slot $T = 1$ for simplicity. We can observe that there is an optimal value of first time slot τ to maximize the EE. In general, with the increase of

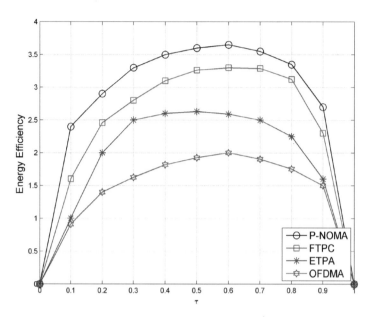

Fig. 9.7 Energy efficiency versus time allocation

time τ, the system EE first increases, then reaches its optimal value and finally decreases. Such phenomenon can be observed for all the cases. In addition, it can be seen that for different algorithms, the optimal EE is different, which evidences the advantages of proposed resource allocation scheme, and the necessity to investigate the time allocation schemes. For example, for the proposed P-NOMA, the optimal EE is higher than the others. The performance difference can be up to 60%, e.g., when comparing the P-NOMA with OFDMA at $\tau = 0.6$. As we can see, our proposed scheme outperforms the other schemes, which confirms the advantages of our proposed schemes.

NOMA is considered as one of the promising techniques for increasing the data rates in the future mobile communication systems. By applying successive interference cancellation schemes and superposition coding at the NOMA receiver, multiple users can be multiplexed on the same subchannel. In this chapter, we have investigated secure-rate and energy-efficient resource allocation problem for NOMA systems empowered by the wireless power transfer. With the explicit consideration of an existing eavesdropper, the objective of this chapter is to provide secure and energy-efficient transmission scheme among multiple users by investigating time, power and subchannel allocation schemes. In order to solve the formulated problem, we propose an iterative algorithm with guaranteed convergence to deliver a competitive suboptimal solution. The algorithm is also capable of providing global optimal solutions for some identified cases. Performance evaluations have demonstrated the effectiveness of the proposed algorithm over other resource allocation schemes in NOMA or OFDMA system.

Chapter 10
Dynamic Computation Offloading Scheme for Fog Computing System with Energy Harvesting Devices

10.1 Framework of Socially Aware Dynamic Computation Offloading for Fog Computing System with EH Devices

In this section, we firstly provide a detailed description of the system model of the considered fog computing system, and then present the formulation of the socially aware dynamic computation offloading problem.

10.1.1 System Movel

(1) System Assumption

As shown in Fig. 10.1, the system consists of N MDs with social relations, fog node and a control cloud. All the MDs are equipped with energy harvesting capabilities. The harvested energy can be stored in the battery and used for local execution or data transmission. Each MD executes an application and generates a series of homogeneous service requests which are independent of each other. Each MD contains one processor, a single server first-in-first-out (FIFO) queue to store arriving requests pending for execution, and the wireless interfaces to connect wireless network. The fog node is deployed to release the load on central cloud and bring low service latency to the nearby users. In this work, we apply queueing theory to model the service latency. Queuing theory has been widely used in the analysis of resource contention in the communication and computing systems, and it is a natural candidate to capture the main features of our systems. In our system, the process queue at the MD is considered as a $M/M/1$ queue, which amounts to assimilate the task arrival process to a Poisson process. The fog node comprises multiple servers, and with a dispatcher that can uniformly distribute the arrival data among the servers. Correspondingly, the queue model of fog node is considered

© Springer Nature Switzerland AG 2021
Z. Zhou et al., *Green Internet of Things (IoT): Energy Efficiency Perspective*,
Wireless Networks, https://doi.org/10.1007/978-3-030-64054-5_10

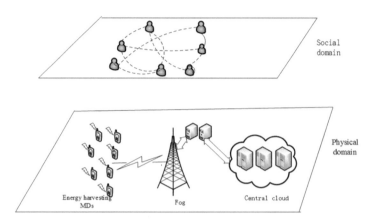

Fig. 10.1 System model

as a $M/G/1$ queue [115]. The central cloud hosts infinite servers in the remote data centers, with no contention among different users, and we model the process as a $M/G/\infty$ queue. We assume that the time is slotted and the length of each time slot is τ. We also denote the time slot and the time slot index set by t and $\mathcal{T} = \{0, 1 \cdots, t \cdots, T - 1\}$.

We assume that the requests generated from MD $i, i \in \mathcal{N}, \mathcal{N} = \{1, 2, \cdots, N\}$ follow a Poisson process with an average arrival rate of $\lambda_{i,t}$ at each time slot t [116]. The requests are assume to be delay sensitive, mutually independent. Each request generating from the MD i contains a data size of θ_i. Each computation request can be executed locally at the MD, or be offloaded to the fog or the central cloud. In addition, none of the above three computation modes may happen, e.g., when the MD does not have sufficient energy, some generated computation requests have to be dropped. The generated requests can be allocated to the local processor, the fog and the central cloud, or even dropped out in parallel at the beginning of the next time slot. So throughout this article, "at time slot t" means the requests are generated at time slot t but executed at time slot $t + 1$. The decision of MD i at time slot t is modeled as a triple $p_{i,t} = \left(p_{i,t}^M, p_{i,t}^F, p_{i,t}^C, p_{i,t}^D \right)$, where $p_{i,t}^M + p_{i,t}^F + p_{i,t}^C + p_{i,t}^D = 1$, which represents the proportion that the requests are executed locally ($p_{i,t}^M$), offloaded to the fog ($p_{i,t}^F$), offloaded to the central cloud ($p_{i,t}^C$), or be dropped ($p_{i,t}^D$) at each time slot t. The MDs compete for the computing resource in the fog in order to minimize their execution cost. They are allowed to make their own strategies without a central authority but can obtain information from a cloud controller. Game theory is employed to model this interaction, where the MDs play with the obtained information from the cloud controller, until they reach a stable state, i.e., a Nash Equilibrium. For ease of reference, we list all the key notations used in our system model in Table 10.1.

Table 10.1 Summary of the key notations

Notations	Meanings
$\mathcal{N}$	The set of energy harvesting MDs
$\mathcal{T}$	The set of time slots
$p_{i,t}^M$	The percentage that MD i executed locally at time slot t
$p_{i,t}^F$	The percentage that MD i offloaded to fog at time slot t
$p_{i,t}^C$	The percentage that MD i offloaded to cloud at time slot t
$p_{i,t}^D$	The percentage that MD i dropped at time slot t
$\boldsymbol{p_{i,t}}$	The decision strategy of MD i at time slot t
$\boldsymbol{p_{i,t}^-}$	The strategies vector other than i at time slot t
$\boldsymbol{p_i}$	Strategies vector of MD i of all the time slots
$\boldsymbol{p_i^-}$	Strategies vector other than MD i of all the time slots
$\boldsymbol{p}$	Strategies vector of all MDs of all the time slots
$\lambda_{i,t}$	The request rate of MD i at time slot t
u_i^M	The computing capability of MD i
$l_{i,t}^M$	The workload of MD i at time slot t
κ_i	The per cycle energy consumption of MD i
θ_i	The data size in each request of MD i
x_i	The CPU cycles required for per bit data of MD i
μ_i	The per rate punishment cost of MD i
W	The channel bandwidth
$q_{i,t}$	The transmission power of MD i at time slot t
$g_{i,t}^{BS}$	The channel gain for MD i at time slot t
$\omega_{i,t}$	The noise power at time slot t
c	The number of servers in the fog
u^F	The server service rate in the fog
l_t^F	The workload of the fog at time slot t
T^{FC}	The fixed delay from fog to the central cloud
u^C	The service rate of the central cloud
$e_{i,t}$	The harvested energy for MD i at time slot t
$B_{i,t}$	The total energy for MD i at time slot t
μ_i	The punishment for per dropped task for MD i
$\bar{\alpha}$	The weight of task dropping punishment

(2) Local Execution Model

Let u_i^M denotes the computing capability of MD i, which is determined by the intrinsic nature of the MD, i.e., CPU Cycle. Different MDs may have different computing capability. Additionally we assume that $l_{i,t}^M$ denotes the normalized workload on the MD i at time slot t, which represents the percentages of CPU that have been occupied. $l_{i,t}^M = 0$ indicates that the CPU is totally idle at time slot

t. When considering a $M/M/1$ queue, the response time of a $M/M/1$ queue is $R = \frac{1/u}{1-\rho}$ [117], where $\rho = \frac{\lambda}{u}$ is the queue utilization, λ is the arrival rate, u is the service rate. Accordingly, the average response time $T_{i,t}^{M}\left(p_{i,t}^{M}\right)$ for locally processing requests at MD i is expressed as follows:

$$
T_{i,t}^{M}\left(p_{i,t}^{M}\right) = \frac{1/u_i^{M}\left(1 - l_{i,t}^{M}\right)}{1 - \frac{\lambda_{i,t}\,p_{i,t}^{M}}{u_i^{M}\left(1 - l_{i,t}^{M}\right)}}
$$

$$
= \frac{1}{u_i^{M}\left(1 - l_{i,t}^{M}\right) - \lambda_{i,t}\,p_{i,t}^{M}}.
$$

(10.1)

Assume that the number of CPU cycles needed for computing 1-bit of input data locally is x_i and the energy consumption per cycle is κ_i for MD i. Then the expression $x_i\kappa_i$ is the per bit computing energy consumption for MD i. x_i and κ_i are related to the intrinsic nature of the CPU and the complexity of the requests for each MD. Then the energy consumption $E_{i,t}^{M}\left(p_{i,t}^{M}\right)$ of local executing the requests for MD i can be given as follows:

$$
E_{i,t}^{M}\left(p_{i,t}^{M}\right) = x_i\kappa_i\,p_{i,t}^{M}\lambda_{i,t}\tau\theta_i.
$$

(10.2)

(3) Fog Execution Model

MD i transmits the data to the fog through a BS at the beginning of the time slot with rational considerations. The wireless channel is assumed to be independent and identically distributed (i.i.d) block fading, i.e. the channel remains static within each time slot, but varies from one to another. Considering the mutual interference caused by other MDs in the system and the non-ignorable background interference, we can obtain the uplink transmission rate for computation offloading of MD i at time slot t as follows [118]:

$$
R_{i,t} = W\log_2\left(1 + \frac{q_{i,t}g_{i,t}^{BS}}{w_{i,t} + \sum_{j\in N, j\neq i} q_{j,t}g_{j,t}^{BS}}\right),
$$

(10.3)

where W is the channel bandwidth, $q_{i,t}$ is the transmission power of the MD i at time slot t, $g_{i,t}^{BS}$ is the channel gain between the MD i and BS at time slot t, $\omega_{i,t}$ is the background noise interference received by MD i at time slot t.

From (10.3), we can then obtain the uplink transmission time $T_{i,t}^{UP}$ of MD i for offloading the data to BS as follows:

$$T_{i,t}^{UP}\left(p_{i,t}^F, p_{i,t}^C\right) = \frac{\left(p_{i,t}^F + p_{i,t}^C\right)\lambda_{i,t}\tau\theta_i}{R_{i,t}}$$

$$= \frac{\left(p_{i,t}^F + p_{i,t}^C\right)\lambda_{i,t}\tau\theta_i}{W\log_2\left(1 + \dfrac{q_{i,t}g_{i,t}^{BS}}{w_{i,t} + \sum_{j\in N, j\neq i} q_{j,t}g_{j,t}^{BS}}\right)}. \tag{10.4}$$

Then, the energy consumption of uplink transmission $E_{i,t}^{UP}\left(p_{i,t}^F, p_{i,t}^C\right)$ can be given as follows:

$$E_{i,t}^{UP}\left(p_{i,t}^F, p_{i,t}^C\right)$$
$$=q_{i,t}T_{i,t}^{UP}\left(p_{i,t}^F, p_{i,t}^C\right)$$
$$= \frac{q_{i,t}\left(p_{i,t}^F + p_{i,t}^C\right)\lambda_{i,t}\tau\theta_i}{W\log_2\left(1 + \dfrac{q_{i,t}g_{i,t}^{BS}}{w_{i,t} + \sum_{j\in N, j\neq i} q_{j,t}g_{j,t}^{BS}}\right)}. \tag{10.5}$$

The fog node is located at the BS and connects to the BS through fiber with large enough bandwidth, so we just neglect the transmission time from the BS to fog node. Accordingly, we assume that there are c homogeneous servers deployed in the fog, and the service rate of each server is denoted as u^F. The requests from different MDs in the system are pooled together with a total rate $\lambda_{total,t}$ at time slot t which also follows Poisson process. Therefore, $\lambda_{total,t}$ is given as follows:

$$\lambda_{total,t} = \sum_{i=1}^{N} p_{i,t}^F \lambda_{i,t}. \tag{10.6}$$

Correspondingly, we assume that the workload of the fog node is denoted as l_t^F $(0 < l_t^F < 1)$, which is the average occupied percentage of each server at time slot t. As a $M/G/1$ queue is considered at the fog node, the average response time at the fog as follows [115, 122]:

$$T_{F,t}\left(p_{i,t}^F\right) = \frac{2u^F\left(1 - l_t^F\right) - \sum_{i=1}^{N} \lambda_{i,t} p_{i,t}^F/c}{2u^F\left(1 - l_t^F\right)\left[u^F\left(1 - l_t^F\right) - \sum_{i=1}^{N} \lambda_{i,t} p_{i,t}^F/c\right]}. \tag{10.7}$$

After the execution is done at the fog node, the results will be delivered to the MDs. And we neglect the time and energy consumption for the MDs to receive the processed requests outcome, due to the fact that for many applications, for example, the face recognition, the size of the computation output in general is much smaller than the size of computation input data [118, 119].

(4) Cloud Execution Model

Additionally, we assume that there is a fixed delay T_{FC} for sending the requests to the central cloud through the fog. As the central cloud has sufficient computing resources to process these requests, the queuing time of the requests in the central cloud can be negligible. The queue model at the central cloud is considered as $M/G/\infty$ with the service rate u^C, which is usually faster than the fog service rate u^F. Then, the response time $T_{C,t}\left(p_{i,t}^C\right)$ of the requests offloaded to the central cloud can be presented as follows:

$$T_{C,t}\left(p_{i,t}^C\right) = \frac{1}{u^C}. \tag{10.8}$$

When the execution at central cloud is done, the results will be delivered to fog and then delivered to the MDs. Similarly, the time and energy consumption for receiving the results for MDs can be neglected.

(5) Energy Harvesting Model

A successive energy packet arrival model is used to model the energy harvesting process. We assume that the arrival of energy packet also follows a Poisson process with an average arrival rate $e_{i,t}$, and $e_{i,t} \leq e_{i,t}^{th}$, we assume that $e_{i,t}^{th}$ is the maximum energy arrival rate, and it is i.i.d in different time slots. The arrived energy will be harvested and stored in the battery, and used for either local execution or computation offloading. Let $B_{i,t}$ denotes the battery energy level for MD i at the beginning of time slot t. Without loss of generality, we assume $B_{i,t} < \infty, \forall t \in T$. In this work, we ignore the energy consumption for other purposes besides local computation and data transmission. Denote the energy consumed by the MD i in time slot t as $E_{i,t}$, which comprises of two parts: (1) energy consumption of local service request processing; (2) energy consumption for sending requests, which depends on the strategy it chooses. We can express $E_{i,t}$ as follows:

$$E_{i,t} = E_{i,t}^M\left(p_{i,t}^M\right) + E_{i,t}^{UP}\left(p_{i,t}^F, p_{i,t}^C\right)$$

$$= x_i \kappa_i p_{i,t}^M \lambda_{i,t} \tau \theta_i + \frac{q_{i,t}\left(p_{i,t}^F + p_{i,t}^C\right)\lambda_{i,t}\tau\theta_i}{W\log_2\left(1 + \frac{q_{i,t}g_{i,t}^{BS}}{w_{i,t}+\sum_{j\in N, j\neq i} q_{j,t}g_{j,t}^{BS}}\right)}. \tag{10.9}$$

$E_{i,t}$ should be smaller than the battery level, i.e.,

$$E_{i,t} \leq B_{i,t} \quad \forall t \in T. \tag{10.10}$$

Thus, the battery energy level of MD i evolves according to the following equation:

$$B_{i,t+1} = B_{i,t} - E_{i,t} + e_{i,t} \quad \forall t \in T. \tag{10.11}$$

It is much more complicated to design the computation offloading policies for the fog computing system with the MDs equipped with energy harvesting function, compared to the conventional MCC systems with battery-powered MDs. Moreover, the system decisions are coupled among different time slots because of the temporally evolved battery energy level. Consequently, it is very challenge to determine the optimal computation offloading strategies, which would balance the computation performances of the current and future computation requests as better as possible.

(6) Social Network Model

In this section, we use a social graph $(\mathcal{N}, \varepsilon^s)$ to denote the social tie structure among the MDs. The vertex set is the N MD members and the edge set is denoted as $\varepsilon^s = \left\{ (i, j) : e^s_{i,j} = 1, \forall i, j \in N \right\}$, where $e^s_{i,j} = 1$ denotes that MD i and j have social relationship between each other, and verse vice. The strength of social relationship between MD i and MD j is denoted as s_{ij}, which is normalized to be $s_{ij} \in [0, 1]$. And the larger the value of s_{ij}, the stronger the social tie between the two MDs. The MD i's social tie to itself is $s_{ii} = 1$. N^S_i is defined as the MD i's social group, which is the set of MDs that have social ties with MD i, i.e. $N^S_i \triangleq \left\{ j \in N \left| e^s_{ij} \in \varepsilon^s \right. \right\}$.

It is worth nothing that the social relationships among MDs can be obtained by locally putting forward the recognition process through the proximity communications technology, for example, the WiFi-Direct and so on, prior to the computation offloading process.

10.1.2 Problem Formulation

In this section, the execution cost is defined as the weighted sum of the execution delay and the task dropping punishment cost, which will be described specifically as follows. The average execution delay for MD i at time slot t can be denoted as follows:

$$T_{i,t}\left(p_{i,t}, p_{i,t}^-\right)$$

$$= p_{i,t}^M T_{i,t}^M\left(p_{i,t}^M\right) + \left(p_{i,t}^F + p_{i,t}^C\right) T_{i,t}^{UP}\left(p_{i,t}^F + p_{i,t}^C\right)$$

$$+ p_{i,t}^F T_{F,t}\left(p_{i,t}, p_{i,t}^-\right) + p_{i,t}^C T_{C,t}\left(p_{i,t}^C\right) \tag{10.12}$$

$$= p_{i,t}^M T_{i,t}^M\left(p_{i,t}^M\right) + p_{i,t}^F\left[T_{i,t}^{UP}\left(p_{i,t}^F + p_{i,t}^C\right) + T_{F,t}\left(p_{i,t}, p_{i,t}^-\right)\right]$$

$$+ p_{i,t}^C\left[T_{i,t}^{UP}\left(p_{i,t}^F + p_{i,t}^C\right) + T_{C,t}\left(p_{i,t}^C\right)\right].$$

where $p_{i,t} = \left(p_{i,t}^M, p_{i,t}^F, p_{i,t}^C, p_{i,t}^D\right)$ is the strategy vector of MD i at time slot t; $p_{i,t}^-$ is the vector formed by the strategies of all MDs except the i-th one at time slot t, which can be denoted as $p_{i,t}^- = \left\{\cdots, p_{i-1,t}^M, p_{i-1,t}^F, p_{i-1,t}^C, p_{i-1,t}^D, p_{i+1,t}^M, p_{i+1,t}^F, p_{i+1,t}^C, p_{i+1,t}^D \cdots\right\}$.

Nevertheless, some of the requests may not be executed but have to be dropped, e.g., due to energy shortage for local computing or offloading to the cloud, meanwhile when the wireless channel from MDs to the fog node is in deep fading, the data of the requests cannot be successfully delivered. To take this aspect into consideration, we penalize per dropped task by cost μ_i, thus the punishment cost for MD i at time slot t can be expressed as follows:

$$C_{i,t} = \mu_i p_{i,t}^D \lambda_{i,t} \tau. \tag{10.13}$$

Consequently, the execution cost for MD i at time slot t, can be formulated as follows:

$$EC_{i,t}\left(p_{i,t}, p_{i,t}^-\right)$$

$$= T_{i,t}\left(p_{i,t}, p_{i,t}^-\right) + \bar{\alpha} C_{i,t}\left(p_{i,t}^D\right)$$

$$= \frac{p_{i,t}^M}{u_i^M\left(1 - l_{i,t}^M\right) - \lambda_{i,t} p_{i,t}^M} + \frac{\left(p_{i,t}^F + p_{i,t}^C\right)^2 \lambda_{i,t} \tau \theta_i}{W \log_2\left(1 + \frac{q_{i,t} g_{i,t}^{BS}}{w_{i,t} + \sum_{j \in N, j \neq i} q_{j,t} g_{j,t}^{BS}}\right)} \tag{10.14}$$

$$+ p_{i,t}^F \frac{2u^F\left(1 - l_t^F\right) - \sum_{i=1}^N \lambda_{i,t} p_{i,t}^F / c}{2u^F\left(1 - l_t^F\right)\left[u^F\left(1 - l_t^F\right) - \sum_{i=1}^N \lambda_{i,t} p_{i,t}^F / c\right]}$$

$$+ p_{i,t}^C\left(T_{FC} + \frac{1}{u^C}\right) + \bar{\alpha} \mu_i p_{i,t}^D \lambda_{i,t} \tau,$$

where $\bar{\alpha}$ is the weight of task dropping cost.

Due to the existence of social relationships, it is natural that users would take into account the effect of its neighbors' decision. So users are coupled in the social domain due to the social ties among them, and MD i aims to choose the strategy $p_{i,t} = \left(p_{i,t}^M, p_{i,t}^F, p_{i,t}^C, p_{i,t}^D\right)$ to minimize its social group execution cost, defined as

$$SEC_{i,t}\left(p_{i,t}, p_{i,t}^-\right) \triangleq EC_{i,t}\left(p_{i,t}, p_{i,t}^-\right) + \sum_{j \in N_i^S} s_{ij} EC_{j,t}\left(p_{j,t}\right). \tag{10.15}$$

So the average social group execution cost of MD i during a period of time can be denoted as follows:

$$MSEC_i\left(p_i, p_i^-\right) = \frac{1}{T}\sum_{t=0}^{T-1} SEC_{i,t}\left(p_{i,t}, p_{i,t}^-\right), \tag{10.16}$$

where $p_i = [p_{i,0}, p_{i,1}, \cdots, p_{i,t}, \cdots]$ is the strategies vector of MD i of all the time slots; p_i^- is the strategies vector of all MDs other than MD i of all the time slots, which can be denoted as $p_i^- = \{\cdots, p_{i-1,0}, p_{i-1,1}, \cdots, p_{i-1,T}, p_{i+1,0}, p_{i+1,1}, \cdots, p_{i+1,T}, \cdots\}$.

We next consider the distributed decision making problem among the MDs for making their average social group execution cost minimize. We formulate the problem as follows:

$$\min_{p_i} \quad MSEC_i\left(p_i, p_i^-\right) \tag{10.17}$$

Subject to

$$\sum_{i=1}^{N} p_{i,t}^F \lambda_{i,t} - cu^F\left(1 - l_t^F\right) < 0, \tag{10.18a}$$

$$\lambda_{i,t} p_{i,t}^M - u_i^M\left(1 - l_{i,t}^M\right) < 0, \tag{10.18b}$$

$$p_{i,t}^M + p_{i,t}^F + p_{i,t}^C + p_{i,t}^D = 1, \tag{10.18c}$$

$$0 \leq p_{i,t}^M, p_{i,t}^F, p_{i,t}^C, p_{i,t}^D \leq 1, \tag{10.18d}$$

$$x_i \kappa_i p_{i,t}^M \lambda_{i,t} \tau \theta_i + \frac{q_{i,t}\left(p_{i,t}^F + p_{i,t}^C\right)\lambda_{i,t}\tau\theta_i}{v_{i,t}} \leq B_{i,t}, \tag{10.18e}$$

$$B_{i,t+1} = B_{i,t} - E_{i,t} + e_{i,t}, \tag{10.18f}$$

$$\forall i \in \mathcal{N}, t \in \mathcal{T}, \tag{10.18g}$$

where $v_{i,t} = W\log_2\left(1 + \frac{q_{i,t}g_{i,t}^{BS}}{w_{i,t}+\sum_{j\in N, j\neq i} q_{j,t}g_{j,t}^{BS}}\right)$. Constraint (10.18a) is derived from (10.7). It shows that the request arrival rate at each server should not exceed the service rate to ensure a stable queue. Nevertheless, due to the energy causality constraints (10.18d) and (10.18e), the MDs' decisions are coupled among different time slots, which makes the problem difficult to be tackled. From [120], we can find that by introducing a non-zero lower bound, $E_{i,t}^{\min}$, and a reasonable upper bound $E_{i,t}^{\max}$, on the battery at each time slot, such coupling effect can be eliminated and the system operation can be optimized by ignoring energy constraints (10.18d) and (10.18e). Thus, we introduce a modified version of the above problem as follows

$$\min_{p_i} \quad MSEC_i\left(p_i, p_i^-\right)$$

Subject to

$$(18a), (18b), (18c), (18g) \tag{10.19a}$$

$$E_{i,t} \in \{0\} \cup \left[E_{i,t}^{\min}, E_{i,t}^{\max}\right] \tag{10.19b}$$

In the following, we propose a game theoretic approach in order to achieve efficient computation offloading decision makings among the MDs. Game theory is a powerful tool to analyze the interactions among multiple users who focus on their own interests. A Nash Equilibrium has the nice self-stability property that all the users can achieve a mutually satisfactory solution and no user has the incentive to deviate unilaterally. Moreover, by using the intelligence of each individual MD user, game theory is a useful framework for devising decentralized mechanisms with low complexity, such that users can self-organize into a mutually satisfactory solution. It can ease the heavy burden of complex centralized management and reduce the communicating overhead between the fog and MDs.

In this setting, the decisions of all MDs are mutually dependent and the proposed model is a GNEP. The GNEP differs from classical Nash Equilibrium Problem (NEP) in that, while in a NEP, only the players' objective functions depend on the other players' strategies, but in a GNEP both the objective functions and the strategy sets depend on the other players' strategies. In our problem, the dependence of each player strategy set on the other players' strategies is represented by the constraint (10.18a), which includes all the MDs' decision variables. Then we will introduce a lemma to illustrate the proposed problem is a jointly convex GNEP so that we can use exponential penalty function and semi-smooth Newton method proposed in the following section to address it effectively.

Lemma 10.1 *The derived GNEP is a jointly convex GNEP.*

Proof The detailed proof is given in [121].   □

10.2 Proposed Solution

In this section, as we have proved that the derived GNEP is a jointly convex problem, we can use exponential penalty function method to solve the GNEP above. We reformulate the GNEP as a sequence of smoothing penalized NEPs by means of a partial penalization of the coupling constraints where the exponential penalty functions are used. Furthermore, we formulate the KKT conditions for smoothing penalized NEPs into a system of non-smooth equations, and then apply the semi-smooth Newton method with Armijo line search to solve the system. The specific algorithm is shown below.

By careful observation, we can find that (10.18a) is a coupled constraint, which includes other participants' decision variables. Other constraints, such as (10.18b), (10.19b) only depend on the MD i itself. As we all know, solving a classic NEP is much easier than solving a GNEP. In this section, through partial punishing the difficult coupling constraints in GNEP, which helps us to solve a classical NEP instead of solving a more difficult GNEP.

With punishing the coupling constraints with exponential penalty function, the original problem is changed as follows:

$$\min_{p_i} \quad MSEC_i\left(p_i, p_i^-\right) + \frac{1}{\rho}\sum_{t=0}^{T-1} \exp\left[\rho g_{i,t}\left(p_{i,t}, p_{i,t}^-\right)\right] \tag{10.20}$$

s.t.

$$\lambda_{i,t}\, p_{i,t}^M - u_i^M\left(1 - l_{i,t}^M\right) < 0 \tag{10.21a}$$

$$x_i\kappa_i\, p_{i,t}^M \lambda_{i,t}\tau\theta_i + \frac{q_{i,t}\left(p_{i,t}^F + p_{i,t}^C\right)\lambda_{i,t}\tau\theta_i}{v_{i,t}} - E_{i,t}^{\max} \leq 0 \tag{10.21b}$$

$$E_{i,t}^{\min} - x_i\kappa_i\, p_{i,t}^M \lambda_{i,t}\tau\theta_i - \frac{q_{i,t}\left(p_{i,t}^F + p_{i,t}^C\right)\lambda_{i,t}\tau\theta_i}{v_{i,t}} \leq 0 \tag{10.21c}$$

$$p_{i,t}^M + p_{i,t}^F + p_{i,t}^C + p_{i,t}^D = 1;\ \forall t \in \mathcal{T} \tag{10.21d}$$

where $g_{i,t}\left(p_{i,t}, p_{i,t}^-\right)$ is the short description of $\sum_{i=1}^{N} p_{i,t}^F \lambda_{i,t} - cu^F\left(1 - l_t^F\right)$. Here, we relax the constraints as we neglect such a constraint denoted by $0 \leq p_{i,t}^M, p_{i,t}^F, p_{i,t}^C, p_{i,t}^D \leq 1$. Nevertheless, the relaxation has no effect on the optimal strategy vector as we can autonomously select the optimal ones ranged on [0, 1] at last.

Then the convergence theorem of the exponential penalty function method is given in the following theorem.

Theorem 10.1 *Let* $\{\rho^k\}$ *be a positive sequence that tends to be infinite. For each k,* p_i^k *is the solution of the NEP when* $\rho = \rho^k$. *Set* $\bar{p}_i$ *is a cluster point of the sequence* $\{p_i^k\}$, *and satisfies the inequality constraints, thus* $\bar{p}_i$ *is the solution of the GNEP.* ☐

Proof See the theorem 1 in Ref. [123]. ☐

So the original GNEPs are evolved to a list of classical NEPs. From Theorem 10.1, we can find that solving a list of NEPs can obtain the solution of GNEP. For the sake of simplicity, we use $h_{i,t}^{(k)}\left(p_{i,t}\right) \leq 0$ $(k = 1, 2, 3, 4)$ to replace the constraint (10.21a), (10.21b), (10.21c), (10.21d) respectively. So the derived classical NEP denoted in a simple form is expressed as following:

$$\min_{p_i} \quad MSEC_i\left(p_i, p_i^-\right) + \frac{1}{\rho} \sum_{t=0}^{T-1} \exp\left[\rho g_{i,t}\left(p_{i,t}, p_{i,t}^-\right)\right]$$

$$\text{s.t.} \quad h_{i,t}^{(k)}\left(p_{i,t}\right) \leq 0, k = 1, 2, 3, 4; t = 0, \cdots T - 1$$

(10.22)

The KKT condition for the NEP displayed in (10.22) can be denoted as follows:

$$\nabla_{p_i} MSEC_i\left(p_i, p_i^-\right) + \sum_{t=0}^{T-1} \exp\left[\rho g_{i,t}\left(p_i\right)\right]\nabla_{p_i} g_{i,t}\left(p_i, p_i^-\right)$$

$$+ \sum_{t=0}^{T-1}\sum_{k=1}^{4} \beta_{i,t}^{(k)}\nabla_{p_i}\left(h_{i,t}^{(k)}\right) = 0$$

(10.23)

Obviously, assemble all of the systems for $i = 1, 2, \cdots, N$, we can obtain the following equivalent system:

$$L(p, \beta) = \left\{ \begin{array}{c} \nabla_{p_i} MSEC_i\left(p_i, p_i^-\right) + \sum_{t=0}^{T-1} \exp\left[\rho g_{i,t}\left(p_i\right)\right] \\ \\ \nabla_{p_i} g_{i,t}\left(p_i\right) + \sum_{t=0}^{T-1}\sum_{k=1}^{4} \beta_{i,t}^{(k)}\nabla_{p_i}\left(h_{i,t}^{(k)}\right) \end{array} \right\}_{i=1}^{N} = 0 \quad (10.24)$$

where $p = [p_1, p_2, \cdots, p_i \cdots, p_N]$ is all the MDs' strategy vector, and $\beta = \left(\beta_{i,t}^{(k)}\right)^T$, $k = 1, 2, 3, 4$; $i = 1, 2, \cdots N$; $t = 1, 2, \cdots T$ is the coefficients of KKT condition for all the MDs during the time slots. We can see that $L(p, \beta)$ is a zero vector of large dimension. To solve p and β, we introduce Fischer–Burmeister (F-B) function $\varphi(a, b) := \sqrt{a^2 + b^2} - (a + b)$, and construct the following form:

$$\Phi(\boldsymbol{\omega}) = \Phi(\boldsymbol{p}, \boldsymbol{\beta}) = \begin{pmatrix} L(\boldsymbol{p}, \boldsymbol{\beta}) \\ \phi(H(\boldsymbol{p}), \boldsymbol{\beta}) \end{pmatrix} = 0 \tag{10.25}$$

where $\phi(H(\boldsymbol{p}), \boldsymbol{\beta}) := \left(\cdots, \varphi\left(-h_{i,t}^{(k)}, \beta_{i,t}^{(k)}\right), \cdots \right)^T$.

We use the semi-smooth Newton method to solve the system $\Phi(\boldsymbol{\omega}) = 0$, which is equal to solving Problem (10.20). Firstly, we define the value function $\Psi(\boldsymbol{\omega}) := \frac{1}{2}\Phi^T(\boldsymbol{\omega})\Phi(\boldsymbol{\omega})$, then Semi-smooth Newton method is described as Algorithm 10.1. At last, we choose the optimal solution on interval $(0, 1)$ from the obtained solutions from Algorithm 10.1.

Algorithm 10.1 Proposed semi-smooth Newton algorithm

1: **Step 0: Initialization**

   Setting initial point: $\boldsymbol{\omega}^0 = (\boldsymbol{p}^0, \boldsymbol{\beta}^0)$, $\rho > 2$, $\sigma \in \left(0, \frac{1}{2}\right)$, $\alpha \in (0, 1)$, $\varepsilon \geq 0$, $k1 \in (0, 1)$,

   $p1 > 2$, let $k = 0$.

2: **Step 1: Termination determination**

3: **if** $\|\Psi(\boldsymbol{\omega}^k)\| \leq \varepsilon$ **then**

4:     Return $\boldsymbol{\omega}^k = (\boldsymbol{p}^k, \boldsymbol{\beta}^k)$;

5: **else**

6:     Go to step 2;

7: **end if**

8: **Step 2: Direction generation**

   Choosing $H_k \in \partial\Phi(\boldsymbol{\omega}^k)$, solving the value of $\boldsymbol{d}^k$ the system

   $H_k\boldsymbol{d}^k = -\Phi(\boldsymbol{\omega}^k)$

9: **if** $\boldsymbol{d}^k$ can't be obtained or does not satisfy the following condition:

   $\nabla\Psi(\boldsymbol{\omega}^k)^T \boldsymbol{d}^k \leq -k1\|\boldsymbol{d}^k\|^{p1}$ **then**

10:     Setting $\boldsymbol{d}^k = -\nabla\Psi(\boldsymbol{\omega}^k)$;

11: **end if**

12: **Step 3: Armijo linear search**

13: Searching the smallest nonnegative integer m^k which satisfies the following inequality:

    $\Psi(\boldsymbol{\omega}^k + \alpha^{m_k}\boldsymbol{d}^k) \leq \Psi(\boldsymbol{\omega}^k) + \sigma\alpha^{m_k}\nabla\Psi(\boldsymbol{\omega}^k)^T \boldsymbol{d}^k$.

14: **Step 4: Rectification**

    Setting $\boldsymbol{\omega}^{k+1} = \boldsymbol{\omega}^k + \alpha^{m_k}\boldsymbol{d}^k$, $k = k + 1$, return to Step 1.

We also show the convergence of the Semi-smooth Newton algorithm in the following Proposition 10.1.

Proposition 10.1 *The Semi-smooth Newton algorithm, with strong system computing power, inherits many excellent features from the classic Newton algorithm. Through determining the step size of the Newton direction with the linear search strategy, it avoids the sensitivity of the algorithm to initial values. Thus the local convergence becomes global convergence.*   □

Proof See the section 5 in Ref. [124].   □

Table 10.2 Simulation parameters

Parameters (Units)	MD 1	MD 2	MD 3
u_i^M (MIPS)	1.6	1.8	2.0
$l_{i,t}^M$	0.10	0.30	0.25
θ_i (bits)	8e+6	8e+6	8e+6
κ_i (joule/cycle)	0.1	0.1	0.1
x_i (cycles)	500	600	700
$E_{i,t}^{max}$ (joules)	10	10	10
$E_{i,t}^{min}$ (joules)	0	0	0

10.3   Performance Evaluation

In this section, extensive simulations are conducted to illustrate the effectiveness of the proposed algorithm for the GNEP. The parameters of the MDs are given as follows in Table 10.2. For simplify, we assume the fog node consists of 10 servers and the service rate of each server is 4 (MIPS), and the workload of each server is $l_t^F = 0.5$. The transmission time from the fog to the central cloud is $T^{FC} = 0.5$ (Second), and the service rate of the central cloud is $u^C = 10$ (MIPS).

The social ties among the three MDs are denoted as a matrix, which is $\begin{bmatrix} 1 & 0.25 & 0.80 \\ 0.50 & 1 & 0.65 \\ 0.55 & 0.90 & 1 \end{bmatrix}$. For this matrix, as we have emphasized in Sect. 10.1.1, the social tie to each MD itself is $s_{ii} = 1$. So we can see that the diagonal elements of the matrix are all 1. Other elements denote the strength of social tie between MD i and MD j, whose values are in interval $(0, 1)$. We can also find that it is a non-symmetric matrix, in other words, the closeness between two MDs is not identical, which is consistent with the relationship between people in our life. For example, the strength of social tie for MD 1 to MD 2 is 0.25, but the strength of social tie for MD 2 to MD 1 is 0.50.

Firstly, we consider the case that there is only one time slot. In this case, we investigate the optimal decision strategy for locally, fog, central cloud execution and dropped out with request arrival rate increasing for MD 2 in Fig. 10.2. We can find that with arrival rate increasing, the proportion of local execution first increases slowly at interval $[0, 1]$, and then increases more faster at interval $[1, 2]$. On the contrary, the proportion of fog execution and central cloud execution decrease slowly at interval $[0, 1]$, and decrease quickly at interval $[1, 2]$. This is because that the MD prefers to offload some requests to the fog when the fog have extra capacity to execute the requests. With all MDs' requests pooling to the fog node with limited resource, each MD has to reduce the proportion of offloading to fog. With the interference from other MDs in the uplink transmission, it would cost quite

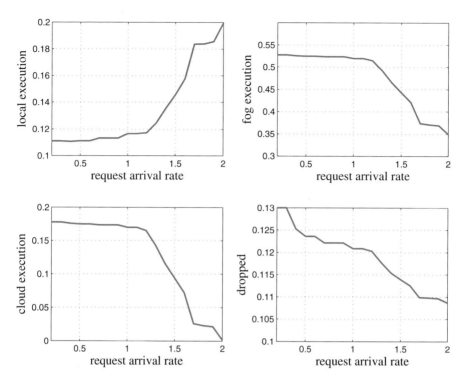

Fig. 10.2 Optimal execution proportion under different request arrival rates

a lot of time and energy to transmit the requests, so the proportion of central cloud execution is also reduced. With the requests arrival rate increasing, the dropped proportion is also reduced.

In Fig. 10.3, we also investigate the impact of request arrival rate on execution delay and energy consumption for local execution, uplink transmission, fog execution, and also the total process. It needs to be emphasized that we neglect the energy consumption of the MD when the requests are executed in the fog, so we can find a straight line in subplot Fig. 10.3. Generally, a larger request arrival rate can result in a larger execution delay and energy consumption, which can be easily found in any curve in Fig. 10.3. Additionally, with the request arrival rate increasing, the consumed energy is also increasing, so we must carefully design the offloading strategy to make the execution cost minimize while meeting the energy supply.

Additionally, we also compare the decision strategy for MD 2 under a certain request arrival rate for different values of ρ. When $\rho = 10$, $\left[p_{2,t}^M, p_{2,t}^F, p_{2,t}^C, p_{2,t}^D \right] = [0.1859, 0.5586, 0.1135, 0.1420]$; when $\rho = 100$, the values are $\left[p_{2,t}^M, p_{2,t}^F, p_{2,t}^C, p_{2,t}^D \right] = [0.2164, 0.5378, 0.1100, 10.1358]$; when $\rho = 1000$, the values are $\left[p_{2,t}^M, p_{2,t}^F, p_{2,t}^C, p_{2,t}^D \right] = [0.2170, 0.5375, 0.1098, 0.1357]$;

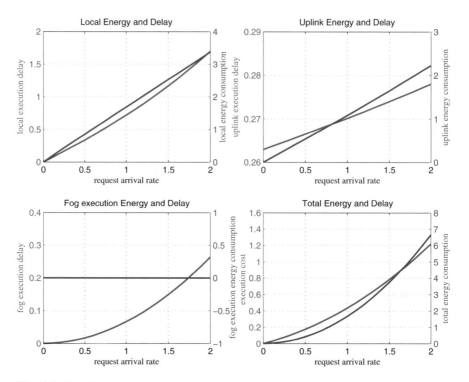

Fig. 10.3 Execution delay and energy consumption under different request arrival rates

when $\rho = 10000$, the values are $\left[p_{2,t}^{M}, p_{2,t}^{F}, p_{2,t}^{C}, p_{2,t}^{D} \right] = [0.2171, 0.5375, 0.1098,$ 0.1356]. The numerical results are displayed in Fig. 10.4. We can see that a larger value of ρ can get a better solution, and the solution would slowly converge to the optimal one with the value of ρ increasing.

Then we focus on a complex situation where the system consists of different time slots. Firstly, we investigate the execution cost among time slots under different request arrival rate, while we assume that in each time slot the harvested energy is identical. It is natural that at the end of each time slot, the remaining energy becomes less, so the MD has to execute more requests locally which would resulting in a larger delay, which means a larger execution cost. We can find this rule from Fig. 10.5. Typically, a larger request arrival rate cause a larger execution cost, which can be observed by comparing the three curves in Fig. 10.5.

Next, we also study the execution cost with the number of MDs under different request arrival rates, which is displayed in Fig. 10.6. We can see that the execution costs are also increasing when the number of MDs increases, which means that the execution delay or the punishment cost become larger. As more and more users compete for resources in the fog with each other, thus intensified the channel interference, reducing the channel transmission rate. So the MD has to execute the request locally or drop them, which leads to a larger execution cost.

In Fig. 10.7, we compare our algorithm with the other existing schemes such as the "Successive Convex Approximation Method" proposed in [125] and the "Lyapunov Optimization-based Dynamic Computation Offloading (LODCO) Algorithm" proposed in [126]. By varying the arrival rates, we can find that our proposed algorithm has a better system performance. This is due to the fact that with the proposed scheme, the GNE point can be achieves, which is a mutually satisfactory solution and no one has the incentive to deviate.

In this chapter, a computation offloading problem in a fog computing system has been investigated. We derive the analytic results of energy consumption, delay performance and cost with assumption of three different queue models at mobile devices, fog and central cloud. By leveraging the obtained results, we take the social relationships of the energy harvesting MDs into the design of offloading scheme. With the objective to minimize the social group execution cost, we advocate game theoretic approach and propose a dynamic computation offloading scheme. Specifically, we formulate a GNEP with various constraints and addressed it by using exponential penalty function method and semi-smooth Newton method with Armijo line search. Extensive performance evaluations are presented to illustrate the effectiveness of the proposed scheme.

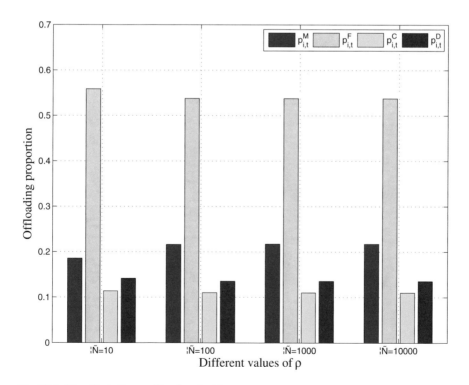

Fig. 10.4 The effect of ρ on offloading decision

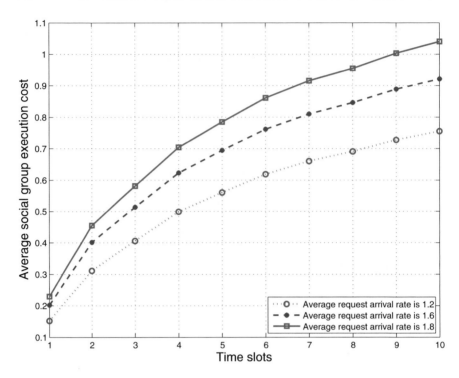

Fig. 10.5 Execution cost among different time slots under different request arrival rates

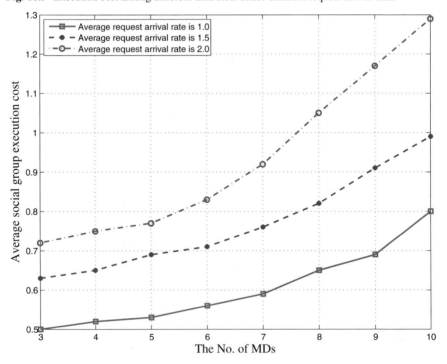

Fig. 10.6 Average execution cost under different request arrival rates

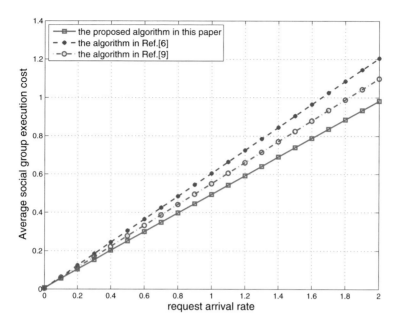

Fig. 10.7 Performance comparison

Chapter 11
Energy-Efficient Resource Allocation for Wireless Powered Massive MIMO System with Imperfect CSI

11.1 Framework of Resource Allocation for Wireless Powered Massive MIMO System with Imperfect CSI

In this section, we firstly provide a detailed description of the system model of wireless powered massive MIMO system with imperfect CSI, analytical models, and then present the formulation of the energy-efficient resource allocation problem.

11.1.1 System Model

As shown in Fig. 11.1, we consider a multi-user massive MIMO system with WPT. In the system, there is one BS, K mobile users and the set of users is denoted by $\mathcal{K}$. The BS is equipped with $N \gg 1$ antennas and each user is equipped with one antenna. In this model, the role of the BS is to charge the users via downlink WPT, while the users have the functionality of storing the energy transmitted by the BS and use the received energy to deliver data to the BS in the uplink. The users can also deliver the CSI through a feedback channel to the BS. For the channel estimation, the BS first sends preambles, and the user performs channel estimation in an interval of symbol periods. Then, the MT feeds the CSI back to the BS. Such a process can be handled in a standardized way. Since the usually CSI feedback is general small comparing with the transmit data, we mainly focus on the energy and data transmission [127].

We assume that the whole transmission process including WPT in the downlink and data transmission in the uplink is within a time block T. In the downlink, BS will use a power P_t transfer energy to all the users and the duration of WPT time τ_k will depend on the individual user k, which should be further optimized. As shown in Fig. 11.2, in the first time slot τ_k, the BS charges user k via WPT and the user

© Springer Nature Switzerland AG 2021
Z. Zhou et al., *Green Internet of Things (IoT): Energy Efficiency Perspective*,
Wireless Networks, https://doi.org/10.1007/978-3-030-64054-5_11

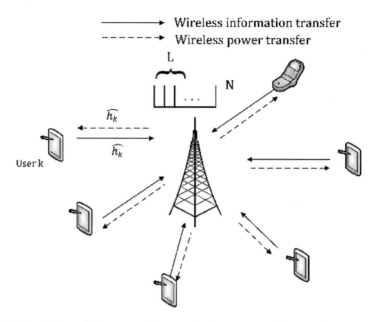

Fig. 11.1 A multi-user wireless powered communications system with transmit antenna selection

downlink:	uplink:
wireless power transfer	wireless information transfer
τ_k	$T - \tau_k$

Fig. 11.2 Time protocol for wireless information and power transfer

stores the harvested energy in a rechargeable battery. Then, in the time duration $T - \tau_k$, user k sends its own data to the BS.

We consider a quasi-static block fading channel model where the channel between the BS and user is constant for a given transmission block T, and can vary independently from one block to another. In each transmission block, user k uses a minimum mean square error (MMSE) channel estimator to estimate the channel. The estimated channel is denoted by $\hat{\mathbf{h}}_k$ and the estimation error is $\hat{\mathbf{e}}_k$. Thus, we have the expression of imperfect CSI as

$$\hat{\mathbf{h}}_k = \mathbf{h}_k + \hat{\mathbf{e}}_k, \tag{11.1}$$

where $\mathbf{h_k}$ is the channel coefficient and we assume $\hat{\mathbf{e}}_k \sim \mathcal{CN}(0, \sigma_{e_k}^2 I_{e_k})$, where I_{e_k} is the identity matrix. The BS is equipped with N antennas, where N is very large. Meanwhile, each antenna of BS requires a separate RF chain, which increases the energy consumption and cost of the massive MIMO system. In order to reduce the energy consumption and improve the system EE, we propose an antenna selection algorithm at the BS, that is, L antennas are selected from the N antennas with the objective to maximize the EE of the considered system. Meanwhile, we also propose to design the energy beamforming vector for the selected antennas to improve the efficiency of WPT. According to the law of conservation of energy, user k can obtain the received energy from the BS as follows [128],

$$E_k = \eta \tau_k (\alpha_k^2 |\mathbf{b}_k^H \mathbf{h}_k|^2 P_t), \tag{11.2}$$

where α_k is the path loss from the BS to user k, $\mathbf{b}_k^H$ is a energy beamforming vector for user k at the BS and $\mathbf{s}$ is the transmitted signal. In the system, when L antennas are selected, we have $\hat{\mathbf{h}}_k \in \mathbb{C}^{1 \times L}$. The transmit power of the BS is P_t. $\eta (0 < \eta \leq 1)$ is the conversion efficiency which transfers the harvested energy into electric energy stored by the user. In order to maximize the harvested energy, we design the energy beamforming policy as $\mathbf{b}_k = \frac{\hat{\mathbf{h}}_k}{\|\hat{\mathbf{h}}_k\|}$ [129], which is named as maximum ratio transmission (MRT) [130]. According to the estimated CSI and beamforming strategy, the energy transfer direction can be adjusted properly to maximize the received energy at the user. Denoting $Q_k = \frac{\sigma_{e_k}^2}{1+\sigma_{e_k}^2} + \frac{\|\hat{\mathbf{h}}_k\|^2}{(1+\sigma_{e_k}^2)^2}$, the obtained energy of user k can be reformed as follows [129]:

$$E_k = \eta \tau_k (\alpha_k^2 Q_k P_t). \tag{11.3}$$

11.1.2   Throughput Analysis

During the second time slot $T - \tau_k$, user k can use the harvested energy to send its data to the BS, and the received signal at the BS is can be expressed as,

$$y_k^{ID} = \sqrt{\frac{E_k}{T - \tau_k}} \alpha_k \hat{\mathbf{h}}_k^H x_k + \mathbf{n}_{u,k}, \tag{11.4}$$

where y_k^{ID} is the received signal at the BS, x_k is the transmitted signal at user k, and $\mathbf{n}_{u,k} \sim \mathcal{CN}(0, \sigma^2)$ is the channel noise. It is also worth noticing that $\frac{E_k}{T-\tau_k}$ is the transmit power of user k.

In a massive MIMO system, with the increase of the number of antennas, the channel hardening effect emerges [131]. Therefore, in order to obtain the expected data rate of the considered system with imperfect CSI, we first study the mutual information distributions with/without antenna selection. To this end, we first arrive at **Theorem** 11.1 about the mutual information distribution of the considered system without antenna selection.

Theorem 11.1 *Given the imperfect CSI and $N \gg 1$ antennas, a numerical approximation of the mutual information in UL of the considered system is given* as,

$$I \sim \mathcal{N}\left(\log_2\left(1 + N\rho_k\right), \frac{(\log_2 e)^2}{N}\right), \tag{11.5}$$

where $\mathcal{N}$ represents standard normal distribution. The signal to interference plus noise ratio (SINR) ρ_k of user k in the uplink can be expressed as:

$$\rho_k = \frac{\frac{E_k \alpha_k^2}{T - \tau_k}}{\sigma^2 + \frac{E_k \alpha_k^2}{T - \tau_k}\sigma_{e_k}^2 + \sum_{j \neq k} \frac{E_j \alpha_j^2}{T - \tau_j}}, \tag{11.6}$$

where σ^2 is noise variance and $\sigma_{e_k}^2$ is the variance of estimation error.

Proof The proof of **Theorem** 11.1 is shown in [132]. □

Theorem 11.1 presents the distribution of mutual information when considering N antennas. Similarly, we can obtain the expression of mutual information when L antennas are selected out of N in **Theorem** 11.2.

Theorem 11.2 *In the considered system, when L antennas are selected, the mutual information distribution is given as follows:*

$$I_{sel} \sim \mathcal{FN}\left(\log_2\left(1 + \left(1 + \ln\frac{N}{L}\right)\rho_k L\right),\right.$$
$$\left.\frac{(\log_2 e)^2 \rho_k^2 L(2 - \frac{L}{N})}{(1 + (1 + \ln\frac{N}{L})\rho_k L)^2}\right), \tag{11.7}$$

where $\mathcal{FN}$ represents the folded normal distribution.

Proof The proof of **Theorem** 11.2 is shown in [132]. □

Obviously, if $L = N$ (N is sufficiently large), the expected value of the distribution is the same as that of the system without antenna selection, and the variance is approximately the same as well. Therefore, adding antenna selection does not affect the channel hardening phenomenon. Thus, in each time block, the expected channel capacity under imperfect CSI is denoted by $E[I]_{im}$:

$$E[I]_{im} = \log_2\left(1 + \left(1 + \ln\frac{N}{L}\right)\rho_k L\right). \tag{11.8}$$

Correspondingly, when L antennas are selected, the throughput is

$$C(P_t, \tau_k, L) = \sum_{k=1}^{K} (T - \tau_k) \log_2 \left(1 + \left(1 + \ln \frac{N}{L} \right) \rho_k L \right). \tag{11.9}$$

Meanwhile, the total energy consumption of the system can be expressed as:

$$U(P_t, \tau_k, L) = P_c \cdot T + P_t \max_{k \in \mathcal{K}} \tau_k, \tag{11.10}$$

where P_c is the constant circuit power consumption, which can be expressed as [109]

$$
\begin{aligned}
P_c &\approx L(P_{DAC} + P_{mix} + P_{filt}) \\
&\quad + K(2P_{syn} + P_{LNA} + P_{mix} + P_{IFA} + P_{filr} + P_{ADC}),
\end{aligned} \tag{11.11}
$$

where $P_{DAC}, P_{mix}, P_{filt}, P_{syn}, P_{LNA}, P_{IFA}, P_{filr}, P_{ADC}$ denotes the power consumption of the DAC, the mixer, the transmit filter, the frequency synthesizer, the low noise amplifier, the frequency amplifier, the receiver filter and ADC, respectively. We denote P_{user} as the power consumption of the each user, i.e., $P_{user} = 2P_{syn} + P_{LNA} + P_{mix} + P_{IFA} + P_{filr} + P_{ADC}$. P_{bs} is expressed as the power consumption for each antenna on the BS, i.e., $P_{bs} = P_{DAC} + P_{mix} + P_{filt}$. Then we have $P_c = KP_{user} + LP_{bs}$. Since the BS and users have to be active for the whole time and the transmit power only exists in the first time slot, in (11.10), the denominator of EE is rewritten as:

$$U(P_t, \tau_k, L) = (KP_{user} + LP_{bs})T + P_t \max_{k \in \mathcal{K}} \tau_k. \tag{11.12}$$

11.1.3 Problem Formulation

With the above analysis, we can obtain the expressions of $C(P_t, \tau_k, L)$ in [bits/s/Hz] and $U(P_t, \tau_k, L)$ in [W]. Correspondingly, the objective of EE in [bits/J/Hz] can be defined as follows,

$$\Pi(P_t, \tau_k, L) = \frac{C(P_t, \tau_k, L)}{U(P_t, \tau_k, L)}. \tag{11.13}$$

From (11.3), (11.9) and (11.13), $\Pi(P_t, \tau_k, L)$ can be given as follows:

$$\Pi(P_t, \tau_k, L)$$

$$= \frac{\sum_{k=1}^{K} (T - \tau_k) \log_2(1 + (1 + \ln \frac{N}{L})\rho_k L)}{(KP_{user} + LP_{bs})T + P_t \max_{k \in \mathcal{K}} \tau_k}. \tag{11.14}$$

With the defined objective, the optimization problem $\mathbf{P_1}$ can be formulated as follows,

$$\max_{P_t, \tau_k, L} \Pi(P_t, \tau_k, L), \tag{11.15}$$

s.t.

$$\mathbf{C1}: \quad 0 \leq P_t \leq P_{bs,max},$$

$$\mathbf{C2}: \quad \frac{E_k}{T - \tau_k} \leq P_{user,max},$$

$$\mathbf{C3}: \quad 0 \leq \tau_k \leq T, \tag{11.16}$$

$$\mathbf{C4}: \quad \frac{C_k}{T - \tau_k} \geq R_{min},$$

$$\mathbf{C5}: \quad L \leq N.$$

In $\mathbf{P_1}$, the objective is to maximize the overall system EE. In (11.16), $\mathbf{C1}$ is the BS transmit power constraint, which shows that the transmit power of the BS cannot be larger than the maximum transmit power $P_{bs,max}$. $\mathbf{C2}$ is the transmit power constraint for user k. $\mathbf{C3}$ means that τ_k cannot be larger than T and $\mathbf{C4}$ can ensure that QoS R_{min} can be meet. Because the channel hardening phenomenon after antenna selection still exists, we can bring (11.3) into $\mathbf{C2}$, to arrive at

$$P_t \leq \frac{P_{user,max}(T - \tau_k)}{\eta \tau_k Q_k \alpha_k^2}. \tag{11.17}$$

Combining $\mathbf{C1}$ and (11.17), we can obtain

$$\tau_k \leq \frac{P_{user,max} T}{(\eta \alpha_k^2 P_{bs,max} Q + P_{user,max})} = \tau_{max}. \tag{11.18}$$

11.2 Proposed Antenna Selection and Resource Allocation Scheme

In this section, antenna selection and resource allocation schemes are introduced to address the formulated problem $\mathbf{P_1}$. At first, we propose an antenna selection scheme to find the optimal number of antennas that the BS can use to obtain EE maximization. Then, power and time allocation schemes are presented to find the optimal transmit power and time duration of WPT.

11.2.1   Proposed Antenna Selection Algorithm

The proposed scheme is based on an improved bisection method to find the solution for antenna selection. The antenna selection scheme is presented in Algorithm 11.1. First, we initialize three variables: the lower bound of the number of antennas, the upper value and the intermediate value, denoted as ω_l, ω_h and ω_m, respectively. Among them, the initial values of ω_l and ω_h are 1 and N, respectively, and the intermediate values are calculated as $\omega_m = \frac{\omega_l + \omega_h}{2}$. In each cycle, we need to compare the two values of $\Pi(\omega_m)$ and $\Pi(\omega_m + 1)$, and determine which subset of the maximum value is located. If $\Pi(\omega_m)$ is less than $\Pi(\omega_m + 1)$, $\omega_m + 1$ is assigned to ω_l; if $\Pi(\omega_m)$ is bigger than $\Pi(\omega_m + 1)$, ω_m is assigned to ω_h. Thus, the maximum value of EE is found by selecting the optimal number of antennas. At the end of each cycle, the ω_m value is updated to the new ω_l or ω_h. When $\omega_h - \omega_l = 1$, the search is ended. Finally, the corresponding L can be obtained.

Algorithm 11.1 Antenna selection algorithm

1: Initialize N, $\Pi(N)$, $\omega_l = 1$, $\omega_h = N$,$\omega_m = \frac{\omega_l + \omega_h}{2}$.
2: **while** $(\omega_h - \omega_l) > 1$ **do**
3:     **if** $\Pi(\omega_m) < \Pi(\omega_m + 1)$ **then**
4:         set $\omega_l = \omega_m + 1$;
5:     **else if** $\Pi(\omega_m) > \Pi(\omega_m + 1)$ **then**
6:         set $\omega_h = \omega_m$;
7:     **else**
8:         break;
9:     **end if**
10: **end while**
11: **if** $\omega_h - \omega_l = 1$ **then**
12:     $\Pi(L) = max\{\Pi(\omega_l), \Pi(\omega_h)\}$;
13: **else**
14:     $\Pi(L) = \Pi(\omega_m)$;
15: **end if**

11.2.2   Power and Time Allocation Schemes

The formulated problem with objective in (11.15) is a non-convex fractional programming problem. Based on the Dinkelbach's method [34], we are able to transform it into a subtractive form. First, given L is obtained, we consider q^* as the global optimal solution of EE, i.e.,

$$q^* = \frac{C(P_t, \tau_k)}{U(P_t, \tau_k)}\big|_{P_t = P_t^*, \tau_k = \tau_k^*}, \tag{11.19}$$

where P_t^* is the optimal transmit power and τ_k^* is the optimal WPT time. Then, we can obtain the following **Theorem** 11.3,

Theorem 11.3 *q can reach its optimal value if and only if*

$$\max_{P_t, \tau_k} C(P_t, \tau_k) - qU(P_t, \tau_k) = 0. \tag{11.20}$$

The proof can be found in [34]. Consequently, problem $\mathbf{P_1}$ can be transformed into a problem $\mathbf{P_2}$:

$$\max_{P_t, \tau_k} \Gamma(P_t, \tau_k), \tag{11.21}$$

s.t.

$$\mathbf{C1, C3, C4,} \tag{11.22}$$
$$\tau_k < \tau_{max},$$

where $\Gamma(P_t, \tau_k) = C(P_t, \tau_k) - q^*U(P_t, \tau_k)$. We can see that $\Gamma(P_t, \tau_k)$ is a concave function with respect to P_t and τ_k as its Hessian matrix is negative semi-define. Therefore, $\mathbf{P_2}$ is now a convex optimization problem and we are able to address it in dual domain to obtain a closed-form solution. The Lagrange dual function corresponding to $\mathbf{P_2}$ is

$$\mathcal{L}(P_t, \tau_k, \alpha, \beta, \mu, \varphi) = C(P_t, \tau_k) - q^*U(P_t, \tau_k)$$
$$- \lambda(P_t - P_{bs,max}) - \beta(\tau_k - \tau_{max}) - \mu(\tau_k - T) \tag{11.23}$$
$$- \varphi(R_{min} - \frac{C_k}{T - \tau_k}),$$

where $\{\lambda, \beta, \mu, \varphi\}$ are the positive Lagrange multipliers associated with the constraint in (11.22), respectively. Correspondingly, the dual problem of (11.23) can be expressed as

$$\mathbf{P_3}: \min_{\alpha, \beta, \mu, \varphi} \max_{P_t, \tau_k} \mathcal{L}(P_t, \tau_k, \alpha, \beta, \mu, \varphi). \tag{11.24}$$

Optimal transmit power P_t^* and the optimal time for WPT τ_k^* can be obtained by solving the KKT condition:

$$\frac{\partial \mathcal{L}(P_t, \tau_k, \lambda, \beta, \mu, \varphi)}{\partial P_t} = 0, \tag{11.25}$$

and

$$\frac{\partial \mathcal{L}(P_t, \tau_k, \lambda, \beta, \mu, \varphi)}{\partial \tau_k} = 0. \tag{11.26}$$

From (11.25), we can obtain

$$P_t^* = \frac{-(\Omega_4 + \Omega_3)\Omega_2 + \sqrt{(\Omega_3 - \Omega_4)^2 \Omega_2^2 + 4 \prod_{i=1}^{5} \Omega_i}}{2\Omega_3 \Omega_4}, \tag{11.27}$$

where $\Omega_1 \sim \Omega_5$ are given as

$$\Omega_1 = \eta \tau_k (\alpha_k^4 Q)(L + L \ln(N/L)),$$

$$\Omega_2 = (T - \tau_K)\sigma^2,$$

$$\Omega_3 = \eta \tau_k (\alpha_k^2 Q)(\sigma_{e_k}^2 + K - 1),$$

$$\Omega_3 = \Omega_1 + \Omega_3,$$

$$\Omega_5 = \frac{(T - \tau_k + \varphi)K}{(\lambda + q^* \max_{k \in \mathcal{K}} \tau_k)(\ln 2)}. \tag{11.28}$$

Next, τ_k^* can be obtained by addressing (11.26) numerically. To obtain the lagrangian multipliers $\lambda, \beta, \mu, \varphi$, the subgradient method with guaranteed convergence [134] can be applied,

$$\lambda(n + 1) = [\lambda(n) - \Delta\lambda(P_{bs,max} - P_t)]^+,$$

$$\beta(n + 1) = [\beta(n) - \Delta\beta(\tau_{max} - \tau_k)]^+,$$

$$\mu(n + 1) = [\mu(n) - \Delta\mu(T - \tau_k)]^+,$$

$$\varphi(n + 1) = [\varphi(n) - \Delta\varphi(\frac{C_k}{T - \tau_k} - R_{min})]^+, \tag{11.29}$$

where n is iteration index, $[x]^+ = max\{0, x\}$, $\Delta\lambda$, $\Delta\beta$, $\Delta\mu$, and $\Delta\varphi$ are the step sizes. Based on the optimal value q^* and the iterative update of the time allocation and power allocation parameter, the convergence can be obtained by satisfying the following relations: $|C(P_t, \tau_k, L) - q^* U(P_t, \tau_k)| < \varepsilon$, where ε is a sufficiently small positive number. If this condition cannot be meet, $q^* = \frac{C(P_t, \tau_k)}{U(P_t, \tau_k)}$ will be updated until the convergence condition is satisfied. The proposed power and time allocation scheme is summarized in Algorithm 11.2.

Algorithm 11.2 Energy efficient resource allocation

1: Initialization:
  $N, L, K, \eta, \alpha_k, P_{bs}, P_{user}, P_{bs,max}, P_{user,max}, R_{min}, \triangle\lambda, \triangle\beta, \triangle\mu,$ and $\triangle\varphi.$
2: Define ε as a sufficiently small positive real number.
3: **while** (!Convergence) **do**
4:     Update $\lambda, \beta, \mu, \varphi$ according to (11.29).
5:     Obtaining the P_t' and τ_k' by solving Eqs. (11.27) and (11.26).
6:     **if** $|C(P_{t'}, \tau_{k'}) - qU(P_{t'}, \tau_{k'})| \leq \varepsilon,$ **then**
7:         Convergence = true,
8:         **return** $P_t^* = P_t', \tau_k^* = \tau_k',$ and obtain optimal q^*
9:     **else**
10:         Convergence = false,
11:         **return** $q = C(P_{t'}, \tau_{k'})/U(P_{t'}, \tau_{k'}),$
12:     **end if**
13: **end while**
14: **return** Obtain P_t^* and $\tau_k^*.$

Table 11.1 Simulation parameters

Parameter	Value
N	100
K	10
$P_{bs,max}$	46 dBm
$P_{user,max}$	23 dBm
R_{min}	0.1 bit/s/Hz
$\triangle\alpha, \triangle\beta, \triangle\mu, \triangle\varphi$	0.001
C	2
η	0.35
ε	0.001
P_{DAC}, P_{ADC}	10 mW
P_{filt}, P_{filr}	2.5 mW
P_{mix}	30.3 mW
P_{syn}	50 mW
P_{LNA}	20 mW
P_{IFA}	3 mW

11.3   Performance Evaluation

In this section, the performance of the proposed scheme is presented and illustrated. Some simulation parameters are given in Table 11.1 [109].

In Fig. 11.3, we present the EE performance of our proposed schemes and prove the effectiveness of the proposed antenna selection and time allocation schemes. To illustrate the advances of the proposed antenna selection scheme, we compare our proposed scheme with the one without antenna selection, which is modified from [133]. It can be clearly seen that the proposed antenna selection scheme can

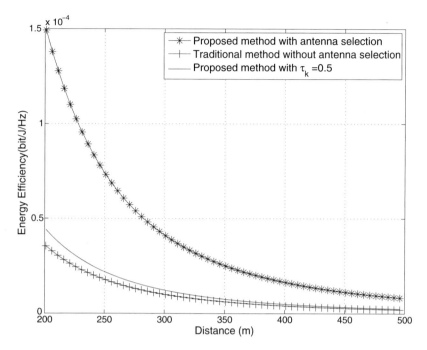

Fig. 11.3 EE w/wo antenna selection and time allocation

improve system EE by selecting the optimal number of antennas from Fig. 11.3. Such observation reveals that the antenna selection has great influence on the EE performance, especially when the BS is close to users. At the same time, we also compare our proposed scheme with the one with equal time allocation, i.e., $\tau_k = 0.5T$ and we assume $T = 1$ for simplicity. Obviously, the proposed algorithm also has the superior performance over the equal time allocation algorithm, which indicates that proper design of time allocation is needed for the SWIPT system. Moreover, It can also be found that as the distance between the BS and users becomes larger, the EE performance decreases.

In Fig. 11.4, we present the impact of CSI imperfection on the system EE performance by varying the variance of channel estimation error $\sigma_{e_k}^2$ and the distance between the BS and the user k. The EE performance of the system with perfect CSI is compared with the EE performance with estimation errors $\sigma_{e_k}^2 = 0.3$ and $\sigma_{e_k}^2 = 0.5$. As we can observe, the system performance with perfect CSI is higher than that with imperfect CSI. When the variance of estimation error increases, the system performance degrades. Also, when the average distance between the BS and users is longer, the performance gap between the one with perfect CSI and the system with the imperfect CSI becomes smaller. For example, when the distance between the BS and the user is 200 m, the system with the perfect CSI has about 4 times higher EE than the one with $\sigma_{e_k}^2 = 0.5$. However, when the distance becomes 450 m, the system with perfect CSI has only 2 times better performance.

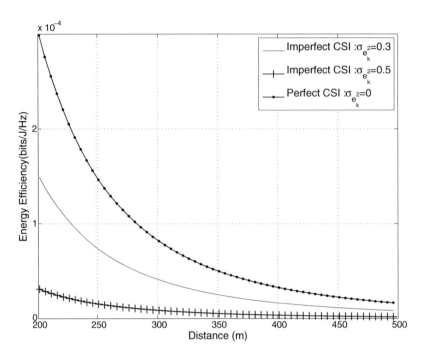

Fig. 11.4 Effect of imperfect CSI and BS-user distance

From Fig. 11.4, the EE of the system when $\sigma_{e_k}^2 = 0.3$ is higher than the one when $\sigma_{e_k}^2 = 0.5$, which confirms that the imperfect CSI has significant impact on the system performance. At the same time, we can also observe that the EE decreases with the increase of the distance between the BS and the users, which is similar to the observation in Fig. 11.3.

Figure 11.5 describes the EE performance when considering a different transmit power with the change of the number of antennas. It can be seen that with the increase of the number of antennas, the EE performance generally first increases and then decreases after reaching the maximum. For the considered system, different transmit power allocation leads to a different optimal number of antennas. For example, when the transmit power is 30 dBm, optimal $L = 30$ and when the transmit power is 35 dBm, $L = 50$. In addition, from the comparison of the EE performance of a different transmit power allocation, we can clearly find that increasing transmit power can not guarantee any increment of EE. Figure 11.5, where the EE of $P_t = 30$ dBm is higher than the other two curves of the EE, also illustrates the effectiveness of the optimized transmit power allocation.

The EE performance for different transmit powers with the change of the number of users K is presented in Fig. 11.6. It can be seen that with the increase of the number of users, the EE performance first increases and then decreases after reaching the maximum. In addition, by comparing the curves, we can clearly see that

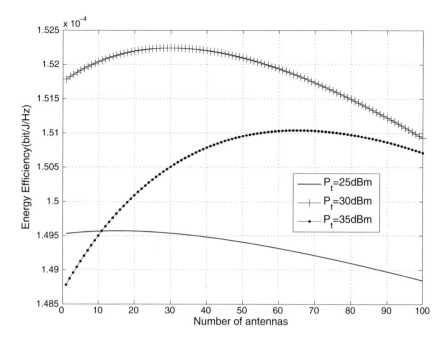

Fig. 11.5 EE vs. number of antennas

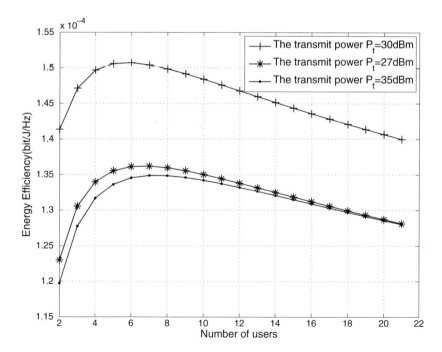

Fig. 11.6 EE vs. number of users

proper increase of transmit power can improve the EE. The EE with $P_t = 30\,\text{dBm}$ is higher than the other two cases, which also illustrates the effectiveness of the proposed transmit power allocation scheme.

In the future wireless network, massive antennas will be explored to improve the system capacity. Meanwhile, as an emerging technique, wireless power transfer offers a potential solution to prolong the lifetime of mobile devices. This chapter studies the energy efficiency of a wireless power transfer enabled multi-user massive MIMO system under imperfect channel estimation. A joint optimization of beamforming design, antenna selection, power and time allocation is studied. In particularly, the antenna selection algorithm is based on an improved bisection scheme to find the optimal number of transmit antennas at the BS. Moreover, a scheme based on nonlinear fractional programming is utilized to address the resource allocation problem and find the optimal power and time allocation. Extensive simulation results can demonstrate the effectiveness of the proposed schemes.

Chapter 12
Summary

In this book, the technologies and applications of green IoT have been studied. Some physical-layer techniques and cross-layer optimization methods on wireless resource allocation and energy management have been proposed and discussed. In the physical layer, the optimization of channel selection, peer discovery, power control, and resource allocation have been researched. The upper-layer requirement cannot be ignored when studying the physical layer. Therefore, the cross layer optimization of relay selection and access control has also been addressed.

In the book, the energy efficiency issues of green IoT have been investigated from the perspectives of D2D, M2M, and air-ground networks. First, an energy-efficient stable matching algorithm has been given for addressing the resource allocation problem in D2D communications, the effectiveness and superiority of which have been validated by extensive simulation results. Then, for M2M communications, a software defined M2M communications network has been proposed to achieve large-scale resource allocation, and an energy-efficient context-aware channel selection algorithm is proposed to achieve small-scale resource allocation. Besides, a long-term cross-layer online resource allocation approach is based on Lyapunov optimization, which jointly optimizes rate control, power allocation, and channel selection without prior knowledge of channel states. Moreover, considering the battery capacity constraints, wireless power transfer and energy harvesting are considered. For energy harvesting enabled energy efficient cognitive M2M communications, we consider the maximizing energy efficiency of M2M-TXs via the joint optimization of channel selection, peer discovery, power control, and time allocation. For OFDM-based NOMA system empowered by wireless power transfer, a novel energy-efficient resource allocation approach with considerations of beamforming design, antenna selection, power allocation, and time division protocol. Finally, the concept of green IoT is extended from single-layer ground network to stereoscopic aerial networks composed of unmanned aerial vehicles, and an energy efficient joint route planning and task assignment approach has been

© Springer Nature Switzerland AG 2021
Z. Zhou et al., *Green Internet of Things (IoT): Energy Efficiency Perspective*,
Wireless Networks, https://doi.org/10.1007/978-3-030-64054-5_12

developed, which can achieve superior performance on energy consumption, utility, and matching satisfaction.

In recent years, green IoT technologies have been in the standardization stage. As a low power wide area (LPWA) technology, the architecture and protocol of narrow band IoT (NB-IoT) architecture and protocol have been standardized in 3GPP Release 13 [135]. Then, wake up signaling (WUS) and early data transmission (EDT) are introduced in Release 15 for optimizing the power consumption and extending battery life [136]. In Release 16 [137], preconfigured uplink resources (PUR) provides the possibility to preassign radio resources to a UE for uplink data transmission without the need of connection setup, which further reduces the power consumption of IoT devices. The IEEE Communication Society has also established a Technical Subcommittee on Green Communications and Computing (TSCGCC) to standardize energy-efficient communications and computing techniques and provide opportunities to collaborative working on solutions for the development of energy-sustainable, resource-saving, and environmentally friendly green IoT. Besides, a growing deal of techniques are standardized and used to implement green IoT, such as discontinuous reception, power saving mode, device clustering, energy harvesting, and power control.

In fact, despite the research and standardization efforts in green IoT, there still exist numerous challenging problems such as the design of unified green IoT architectures, complication of new-generation green infrastructure deployment, utilization of new energy sources, and optimization of energy-efficient communication and computing protocols, all of which need to be investigated in the future work. Firstly, a unified green IoT architecture is essential to develop generic solutions for energy saving and communication across various applications and heterogeneous networks with massive devices. Secondly, to reduce the complexity of new-generation green infrastructure deployment while efficiently exploiting the current infrastructure is less investigated and requires further attention. Thirdly, ambient energy is not always available due to the various nature and human factors, which makes it difficult to reliably support IoT device functionalities without precise energy availability prediction. Finally, existing green techniques rely on complicated optimization algorithms, the complexity of which increase dramatically with the number of IoT devices. Particularly, the complexity is prohibitive considering the massive connection in practice. Low-complexity solutions are urgently required to improve the performance of green IoT, which will play an important role in the next generation wireless communication system.

References

1. Cisco: Cisco Global Cloud Index: Forecast and Methodology, 2016–2021 White Paper. February 2018
2. Zhou, Z., Liao, H., Gu, B., Huq, K.M.S., Mumtaz, S., Rodriguez, J.: Robust mobile crowd sensing: when deep learning meets edge computing. IEEE Netw. **33**(4), 54–60 (2018)
3. Shaikh, F.K., Zeadally, S., Exposito, E.: Enabling technologies for green internet of things. IEEE Syst. J. **11**(2), 983–994 (2017)
4. Said, O., Al-Makhadmeh, Z., Tolba, A.: EMS: An energy management scheme for green IoT environments. IEEE Access **8**, 44983–44998 (2020)
5. Abedin, S.F., Alam, M.G.R., Haw, R., Hong, C.S.: A system model for energy efficient green-IoT network. In: Proceedings of IEEE International Conference on Information Networking, Cambodia, 2015
6. Arshad, R., Zahoor, S., Shah, M.A., Wahid, A., Yu, H.: Green IoT: An investigation on energy saving practices for 2020 and beyond. IEEE Access **5**, 15667–15681 (2017)
7. Wang, R., Hu, H., Yang, X.: Potentials and challenges of C-RAN supporting multi-RATs toward 5G mobile networks. IEEE Access **2**(8), 1187–1194 (2014)
8. Bello, O., Zeadally, S.: Intelligent device-to-device communications in the Internet of Things. IEEE Syst. J. **10**(3), 1172–1182 (2016)
9. Zhang, S., Zhang, N., Zhou, S., Gong, J., Niu, Z., Shen, X.: Energy-aware traffic offloading for green heterogeneous networks. IEEE J. Sel. Areas Commun. **34**(5), 1116–1129 (2016)
10. Zhang, N., Liang, H., Cheng, N., Tang, Y., Mark, J.W., Shen, X.: Dynamic spectrum access in multi-channel cognitive radio networks. IEEE J. Sel. Areas Commun. **32**(11), 2053–2064 (2014)
11. Doppler, K., Rinne, M., Wijting, C., Ribeiro, C.B.: Device-to-device communication as an underlay to LTE-Advanced networks. IEEE Commun. Mag. **47**(12), 42–49 (2009)
12. Tehrani, M.N., Uysal, M., Yanikomeroglu, H.: Device-to-device communications in 5G cellular networks: challenges, solutions, and future directions. IEEE Commun. Mag. **52**(5), 86–92 (2014)
13. Liu, J., Kawamoto, Y., Nishiyama, H., Kato, N., Kadowaki, N.: Device-to-device communications achieve efficient load balancing in LTE-Advanced networks. IEEE Wirel. Commun. Mag. **21**(2), 57–65 (2014)
14. Sheng, M., Li, Y., Wang, X., Li, J., Shi, Y.: Energy efficiency and delay tradeoff in device-to-device communications underlaying cellular networks. IEEE J. Sel. Areas Commun. **34**(1), 92–106 (2016)

© Springer Nature Switzerland AG 2021

Z. Zhou et al., *Green Internet of Things (IoT): Energy Efficiency Perspective*, Wireless Networks, https://doi.org/10.1007/978-3-030-64054-5

15. Peng, M., Yu, Y., Xiang, H., Poor, H.V.: Energy-efficient resource allocation optimization for multimedia heterogeneous cloud radio access networks. IEEE Trans. Multimedia. **18**(5), 879–892 (2016)
16. Babun, L., Yürekli, A.İ., Güvenç, İ.: Multi-hop and D2D communications for extending coverage in public safety scenarios. In: Proceedings of IEEE 40th Local Computer Networks Conference Workshops, pp. 912–919, Clearwater Beach, FL, October. 2015
17. Lohan, E., Koivisto, M., Galinina, O., Andreev, S., Tolli, A., Destino, G., Costa, M., Leppanen, K.: Benefits of positioning-aided communication technology in high-frequency industrial IoT. IEEE Commun. Mag. **56**(12), 142–148 (2018)
18. Ali, A., Feng, L., Bashir, A., El-Sappagh, S.A., Ahmed, S., Iqbal, M., Raja, G.: Quality of service provisioning for heterogeneous services in cognitive radio-enabled Internet of Things. IEEE Trans. Netw. Sci. Eng. **pp**(9), 1–15 (2018)
19. Musaddiq, A., Zikria, Y., Hahm, O., Yu, H., Bashir, A.K., Kim, S.: A survey on resource management in IoT operating systems. IEEE Access. **6**, 8459–8482 (2018)
20. Zhou, Z., Gong, J., He, Y., Zhang, Y.: Software defined machine-to-machine communication for smart energy management. IEEE Commun. Mag. **55**(10), 52–60, (2017)
21. Zhang, C., Zhou, Z., Liu, P., Gu, B.: Resource allocation for energy harvesting based cognitive machine-to-machine communications. In: Proc. IEEE ICC Workshops '19, Shanghai, China, May 2019, pp. 1–6
22. Rico-Alvarino, A., Vajapeyam, M., Xu, H., Wang, X., Blankenship, Y., Bergman, J., Tirronen, T., Yavuz, E.: An overview of 3GPP enhancements on machine to machine communications. IEEE Commun. Mag. **54**(6), 14–21 (2016)
23. Yang, Z., Xu, W., Pan, Y., Pan, C., Chen, M.: Energy efficient resource allocation in machine-to-machine communications with multiple access and energy harvesting for IoT. IEEE Internet Things J. **5**(1), 229–245 (2018)
24. Li, M., Yu, F.R., Si, P., Yao, H., Sun, E., Zhang, Y.: Energy-efficient M2M communications with mobile edge computing in virtualized cellular networks. In: Proceedings of IEEE International Conference on Communications, Paris, France, 2017
25. Yang, Z., Xu, W., Pan, Y., Pan, C., Chen, M.: Energy efficient resource allocation in machine-to-machine communications with multiple access and energy harvesting for IoT. IEEE Internet Things J. **5**(1), 229–245 (2018)
26. Wu, D., Wang, J., Hu, R.Q., Cai, Y., Zhou, L.: Energy-efficient resource sharing for mobile device-to-device multimedia communications. IEEE Trans. Veh. Technol. **63**(5), 2093–2103 (2014)
27. Fodor, G., Dahlman, E., Mildh, G., Parkvall, S., Reider, N., Miklós, G., Turanyi, Z.: Design aspects of network assisted device-to-device communications. IEEE Commun. Mag. **50**(3), 170–177 (2012)
28. Wei, L., Hu, R.Q., Cai, Y., Wu, G.: Delay-optimal dynamic mode selection and resource allocation in device-to-device communications - part I: optimal policy. IEEE Trans. Veh. Technol. **65**(5), 3474–3490 (2016)
29. Asadi, A., Wang, Q., Mancuso, V.: A survey on device-to-device communication in cellular networks. IEEE Commun. Surveys Tut. **16**(4), 1801–1819 (2015)
30. Feng, D., Lu, L., Yi, Y., Li, G.Y., Feng, G., Li, S.: Device-to-device communications underlaying cellular networks. IEEE Trans. Commun. **61**(8), 3541–3551 (2013)
31. Kwon, H., Birdsall, T.: Channel capacity in bits per joule. IEEE J. Ocean. Eng. **11**(1), 97–99 (1986)
32. Roth, A.E., Sotomayor, M.: Two Sided Matching: A Study in Game-Theoretic Modeling and Analysis. Cambridge University Press, Cambridge, UK (1991)
33. Osborne, M.J., Rubinstein, A.: A Course in Game Theory. MIT Press, Cambridge, MA, USA (1994)
34. Dinkelbach, W.: On nonlinear fractional programming. Manag. Sci. **13**(7), 492–498 (1967)
35. Boyd, S., Vandenberghe, L.: Convex Optimization. Cambridge University Press, Cambridge, UK (2004)

36. Cheung, K.T.K., Yang, S., Hanzo, L.: Achieving maximum energy-efficiency in multi-relay OFDMA cellular networks: a fractional programming approach. IEEE Trans. Commun. **61**(8), 2746–2757 (2013)

37. Gale, D., Shapley, L.S.: College admissions and the stability of marriage. Am. Math. Mon. **69**(1), 9–15 (1962)

38. Zhou, Z., Dong, M., Ota, K., Shi, R., Liu, Z., Sato, T.: A game-theoretic approach to energy-efficient resource allocation in device-to-device underlay communications. IET Commun. **9**(3), 375–385 (2015)

39. Miao, G., Himayat, N., Li, G.Y.: Energy-efficient link adaptation in frequency-selective channels. IEEE Trans. Commun. **58**(2), 545–554 (2010)

40. Goldsmith, A.: Wireless Communications. Cambridge University Press, Cambridge, UK (2005)

41. Wang, F., Xu, C., Song, L., Zhao, Q., Wang, X., Han, Z.: Energy-aware resource allocation for device-to-device underlay communication. In: Proceedings of IEEE Communications Conference, pp. 6076–6080 (2013)

42. Joda, R., Lahouti, F., Erkip, E.: Distortion-power tradeoffs in quasi-stationary source transmission over delay and buffer limited block fading channels. IEEE Trans. Wireless Commun. **15**(7), 4505–4520 (2016)

43. Ha, T., Kim, J., Chung, J.: HE-MAC: Harvest-then-transmit based modified EDCF MAC protocol for wireless powered sensor networks. IEEE Trans. Wireless Commun. **17**(1), 3–16 (2018)

44. Saleem, U., Jangsher, S., Qureshi , H., Hassan, S.: Joint subcarrier and power allocation in the energy-harvesting-aided D2D communication. IEEE Trans. Ind. Inf. **14**(6), 2608–2617 (2018)

45. Shih, M.j., Lin, G.Y., Wei, H.Y.: Two paradigms in cellular Internet-of-Things access for energy-harvesting machine-to-machine devices: push-based versus pull-based. IET Wireless Sensor Syst. **6**(4), 121–129 (2016)

46. Wang, H., Ding, G., Wang, J., Wang, L., Taiftsis, T.A., Sharma, P.K.: Resource allocation for energy harvesting-powered D2D communications underlaying cellular networks. In: Proceedings of IEEE International Conference on Communications, Paris, May 2017

47. Zeng, M., Luo, Y., Guo, Q., Jiang, H.: Power allocation for energy harvesting-based D2D communication underlaying cellular network. In: Proceedings of Chinese Control Conference, Dalian, July 2017

48. Aijaz, A., Tshangini, M., Nakhai, M.R., Chu, X. Aghvami, A.: Energy-efficient uplink resource allocation in LTE networks with M2M/H2H co-existence under statistical QoS guarantees. IEEE Trans. Commun. **62**(7), 2353–2365 (2014)

49. Ghavimi, F., Lu, Y., Chen, H.: Uplink scheduling and power allocation for M2M communications in SC-FDMA-based LTE-A networks with QoS guarantees. IEEE Trans. Veh. Technol. **66**(7), 6160–6170 (2017)

50. Zhou, Z., Xiong, F., Xu, C., He, Y., Mumtaz, S.: Energy-efficient vehicular heterogeneous networks for green cities. IEEE Trans. Ind. Informat. **14**(4), 1522–1531 (2018)

51. Luo, Y., Hong, P., Su, R., Xue, K.: Resource allocation for energy harvesting-powered D2D communication underlaying cellular networks. IEEE Trans. Veh. Technol. **66**(11), 10486–10498 (2017)

52. Li, S., Ni, Q., Sun, Y., Min, G., Rubaye, S.A.: Energy-efficient resource allocation for industrial cyber-physical IoT systems in 5G era. IEEE Trans. Ind. Inf. **14**(6), 2618–2628 (2018)

53. Trakhtenbrot, B.A.: A survey of Russian approaches to perebor (brute-force serches) algorithms. Ann. History Comput. **6**(4), 121–129 (1984)

54. Zhou, Z., Guo, Y., He, Y., Zhao, X., Bazzi, W.M.: Access control and resource allocation for M2M communications in industrial automation. IEEE Tran. Ind. Inf. **15**(5), 3093–3103 (2019)

55. Xu, W., Zhou, X., Lee, C., Feng, Z., Lin, J.: Energy-efficient joint sensing duration, detection threshold, and power allocation optimization in cognitive OFDM systems. IEEE Tran. Wireless Commun. **15**(12), 8339–8352 (2016)
56. Sentelle, C., Anagnostopoulos, G.C., Georgiopoulos, M.: Efficient revised simplex method for SVM training. IEEE Trans. Neural Netw. **22**(10), 1650–1661 (2011)
57. Zhou, Z., Ota, K., Dong, M., Xu, C.: Energy-efficient matching for resource allocation in D2D enabled cellular networks. IEEE Trans. Veh. Technol. **66**(6), 5256–5268 (2017)
58. Pei, L., Yang, Z., Pan, C., Huang, W., Chen, M., Elkashlan, M., Nallanathan, A.: Energy-efficient D2D communications underlaying NOMA-based networks with energy harvesting. IEEE Commun. Lett. **22**(5), 914–917 (2018)
59. Zhou, Z., Feng, J., Gu, B., Ai, B., Mumtaz, S., Rodriguez, J., Guizani, M.: When mobile crowd sensing meets UAV: energy-efficient task assignment and route planning. IEEE Trans. Commun. **66**(11), 5526–5538 (2018)
60. Zhou, Z., Sun, C., Shi, R., Chang, Z., Zhou, S., Li, Y.: Robust energy scheduling in vehicle-to-grid networks. IEEE Netw. **31**(2), 30–37 (2017)
61. Zhou, Z., Xiong, F., Huang, B., Xu, C., Jiao, R., Liao, B., Yin, Z., Li, J.: Game-theoretical energy management for energy Internet with big data-based renewable power forecasting. IEEE Access. **5**, 1–14 (2017)
62. Gong, J., Zhou, S., Zhou, Z., Niu, Z.: Policy optimization for content push via energy harvesting small cells in heterogeneous networks. IEEE Trans. Wireless Commun. **16**(2), 717–729 (2016)
63. Meng, Z., Wu, Z., Muvianto, C., Gray, J.: a data-oriented M2M messaging mechanism for industrial IoT applications. IEEE Internet Things J. **4**(1), 236–246 (2017)
64. Duan, L., Gao, L., Huang, J.: Cooperative spectrum sharing: A contract-based approach. IEEE Trans. Intell. Transp. Syst. **13**(1), 174–187 (2014)
65. Liu, T., Li, J., Shu, F., Tao, M., Chen, W., Han, Z.: Design of contract-based trading mechanism for a small-cell caching system. IEEE Trans. Wireless Commun. **16**(10), 6602–6617 (2017)
66. Joshi, S.K., Manosha, K., Codreanu, M., Latva, M.: Dynamic inter-operator spectrum sharing via Lyapunov optimizations. IEEE Trans. Wireless Commun. **16**(10), 6365–6381 (2017)
67. Baki, A.: Continuous monitoring of smart grid devices through multi protocol label switching. IEEE Trans. Smart Grid. **5**(3), 1210–1215 (2014)
68. Neely, M.J.: Stochastic Network Optimization with Application to Communication and Queueing Systems. Morgan and Claypool, USA (2010)
69. Bao, W., Chen, H., Li, Y., Vucetic, B.: Joint rate control and power allocation for non-orthogonal multiple access systems. IEEE J. Sel. Areas Commun. **35**(12), 2798–2811 (2017)
70. Guo, Y., Yang, Q., Kwak, K.S.: Quality-oriented rate control and resource allocation in time-varying OFDMA networks. IEEE Trans. Veh. Technol. **66**(3), 2324–2338 (2017)
71. Peng, M., Yu, Y., Xiang, H., Poor, H.V.: Energy-efficient resource allocation optimization for multimedia heterogeneous cloud radio access networks. IEEE Trans. Veh. Technol. **18**(5), 879–892 (2016)
72. Mao, Y., Zhang, J., Li, Y., Letaief, K.B.: A Lyapunov optimization approach for green cellular networks with hybrid energy supplies. IEEE J. Sel. Areas Commun. **33**(12), 2463–2477 (2015)
73. Qiu, C., Hu, Y., Chen, Y., Zeng, B.: Lyapunov optimization for energy harvesting wireless sensor communications. IEEE Internet Things J. **5**(3), 1947–1956 (2018)
74. Tang, L., Wei, Y., Chen, W., Chen, Q.: Delay-aware dynamic resource allocation and ABS configuration algorithm in hetNets based on Lyapunov optimization. IEEE Access **5**, 23764–23775 (2017)
75. Oh, S.H., Li, K.: BER performance of BPSK receivers over two-wave with diffuse power fading channels. IEEE Trans. Wireless Commun. **4**(4), 321–354 (2005)
76. Liu, C., Bennis, M., Debbah, M., Poor, H.V.: Dynamic task offloading and resource allocation for ultra-reliable low-latency edge computing. IEEE Trans. Commun. **67**(6), 4132–4150 (2019)

77. Ko, S.-W., Han, K., Huang, K.: Wireless networks for mobile edge computing: spatial modeling and latency analysis. IEEE Trans. Wireless Commun. **17**(8), 5225–5240 (2018)
78. Mach, P., Becvar, Z.: Mobile edge computing: a survey on architecture and computation offloading. IEEE Commun. Surveys Tuts. **19**(3), 1628–1656 (2017)
79. You, C., Huang, K., Chae, H., Kim, B.: Energy-efficient resource allocation for mobile-edge computation offloading. IEEE Trans. Wireless Commun. **16**(3), 1397–1411 (2017)
80. Liu, X., Jia, M., Zhang, X., Lu, W.: A novel multichannel Internet of Things based on dynamic spectrum sharing in 5G communication. IEEE Internet Things J. **6**(4), 5962–5970 (2019)
81. Li, Y., Yin, Q., Sun, L., Chen, H., Wang, H.: A channel quality metric in opportunistic selection with outdated CSI over nakagami-m fading channels. IEEE Trans. Veh. Technol. **61**(3), 1427–1432 (2012)
82. Lakshminarayana, S., Assaad, M., Debbah, M.: Transmit power minimization in small cell networks under time average QoS constraints. IEEE J. Sel. Areas Commun. **33**(10), 2087–2103 (2015)
83. Neely, M.: Energy optimal control for time-varying wireless networks. IEEE Trans. Inf. Theory **52**(7), 2915–2934 (2006)
84. Sutton, R., Barto, A.: Reinforcement Learning: A Introduction. MIT Press, Cambridge, MA, USA (2018)
85. Auer, P., Cesa-Bianchi, N., Fischer, P.: Finite-time analysis of the multiarmed bandit problem. Mach. Learn. **47**(2-3), 235–256 (2002)
86. Zhou, Z., Gong, J., He, Y., Zhang, Y.: Software defined machine-to-machine communication for smart energy management. IEEE Commun. Mag. **55**(10), 52–60 (2017)
87. Liao, H., Zhou, Z., Zhao, X., Zhang, L., Mumtaz, S., Jolfaei, A., Ahmed, S. H., Bashir, A. K.: Learning-based context-aware resource allocation for edge-computing-empowered industrial IoT. IEEE Internet Things J. **7**(5), 4260–4277 (2020)
88. Zhou, Z., Jia, Y., Chen, F., Tsang, K., Liu, G., Han, Z.: Unlicensed spectrum sharing: From coexistence to convergence. IEEE Wireless Commun. **24**(5), 94–101 (2017)
89. Sun, Y., Zhou, S., Xu, J.: EMM: energy-aware mobility management for mobile edge computing in ultra dense networks. IEEE J. Sel. Areas Commun. **35**(11), 2637–2646 (2017)
90. Auer, P., Cesa-Bianchi, N., Freund, Y., Schapire, E.: The nonstochastic multiarmed bandit problem. SIAM J. Comput. **32**(1), 48–77 (2002)
91. Zhang, H., Xiao, Y., Cai, L.X., Niyato, D., Song, L., Han, Z.: A hierarchical game approach for multi-operator spectrum sharing in LTE unlicensed. In: Proceedings of IEEE Global Communications Conference, San Diego, December 2015
92. Guan, Z., Melodia, T.: CU-LTE: Spectrally-efficient and fair coexistence between LTE and Wi-Fi in unlicensed bands. In: IEEE International Conference on Computer Communications, San Francisco, April 2016
93. Marcastel, A., Belmega, V., Mertikopoulos, P., Fijalkow, I.: Online power optimization in feedback-limited, dynamic and unpredictable IoT networks. IEEE Trans. Signal Process. **67**(11), 2987–3000 (2019)
94. Modi, N., Mary, P., Moy, C.: QoS driven channel selection algorithm for cognitive radio network: multi-user multi-armed bandit approach. IEEE Trans. Cogn. Commun. Netw. **3**(1), 49–66 (2017)
95. Ali, S., Ferdowsi, A., Saad, W., Rajatheva, N.: Sleeping multi-armed bandits for fast uplink grant allocation in machine type communications. In: Proceedings of IEEE Global Communications Conference, Abu Dhabi, United Arab Emirates, United Arab Emirates, December 2018
96. Paredes, J., Saito, C., Abarca, M., Cuellar, F.: Study of effects of high-altitude environments on multicopter and fixed-wing UAVs' energy consumption and flight time. In: Proceedings of the IEEE Conference on Automation Science and Engineering, August 2017
97. Karypis, G., Kumar, V.: Multilevel graph partitioning schemes. In: Proceedings of the 6th SIAM Conference Parallel Processing for Scientific Computing, Oconomowoc, August 1995
98. Zhou, Z., Zhang, C., Xu, C., Xiong, F., Zhang, Y., Umer, T.: Energy-efficient industrial internet of UAVs for power line inspection in smart grid. IEEE Trans. Ind. Inf. **14**(6), 2705–2714 (2018)

99. Zeng, Y., Zhang, R.: Energy-efficient UAV communication with trajectory optimization. IEEE Trans. Wireless Commun. **16**(6), 3747–3760 (2017)

100. Baker, B., Ayechew, M.: A genetic algorithm for the vehicle routing problem. Comput. Oper. Res. **30**(5), 787–800 (2003)

101. Chu, P.C., Beasley, J.E.: Constraint handling in genetic algorithms: the set partitioning problem. J. Heuristics. **4**(4), 323–357 (1998)

102. Zhong, L., Luo, Q., Wen, D., Qiao, S., Shi, J., Zhang, W.: A task assignment algorithm for multiple aerial vehicles to attack targets with dynamic values. IEEE Trans. Intell. Transp. Syst. **14**(1), 236–248 (2013)

103. Wang, C., Li, J., Ye, F., Yang, Y.: A mobile data gathering rramework for wireless rechargeable sensor networks with vehicle movement costs and capacity constraints. IEEE Trans. Comput. **65**(8), 2411–2427 (2016)

104. Zhou, Z., Gao, C., Xu, C., Zhang, Y., Mumtaz, S., Rodriguez, J.: Social big-data-based content dissemination in internet of vehicles. IEEE Trans. Ind. Informat. **14**(2), 768–777 (2018)

105. Zhang, R., Ho, C.K.: MIMO broadcasting for simultaneous wireless information and power transfer. IEEE Wirel. Commun. **12**(5), 1989–2001 (2013)

106. Zhou, X., McKay, M.R.: Secure transmission with artificial noise over fading channels: achievable rate and optimal rower allocation. IEEE Trans. Veh. Technol. **59**(8), 3831–3842 (2010)

107. Mollanoori, M., Ghaderi, M.: Uplink scheduling in wireless networks with successive interference cancellation. IEEE Trans. Mob. Comput. **13**(5), 1132–1144 (2014)

108. Diamantoulakis, P., Pappi, K., Ding, Z., Karagiannidis, G.: Wireless powered communications with non-orthogonal multiple access. IEEE Trans. Wirel. Commun. **15**(12), 8422–8436 (2016)

109. Cui, S., Goldsmith, A.J., Bahai, A.: Energy-efficiency of MIMO and cooperative MIMO techniques in sensor networks. IEEE J. Sel. Areas Commun. **22**(6), 1089–1098 (2004)

110. Lei, L., Yuan, D., V, P.: On power minimization for non-orthogonal multiple access (NOMA). IEEE Commun. Lett. **20**(12), 2458–2461 (2016)

111. Chang, Z., Gong, J., Ristaniemi, T., Niu, Z.: Energy efficient resource allocation and user scheduling for collaborative mobile clouds with hybrid receivers. IEEE Trans. Veh. Technol. **65**(12), 9834–9846 (2016)

112. Saito, Y., Kishiyama, Y., Benjebbour, A., Nakamura, T., Li, A.: Non-orthogonal multiple access (NOMA) for cellular future radio access. In: Proceedings of IEEE 77th Vehicular Technology Conference, Dresden, June 2013

113. Ng, D.W.K., Lo, E.S., Schover, R.: Energy-efficient resource allocation for secure OFDMA systems. IEEE Trans. Veh. Technol. **61**(6), 2572–2585 (2012)

114. Chang, Z., Lei, L., Zhang, H., Risaniemi, T., Chatzinotas, S., Ottersten, B., Han, Z.: Energy-efficient and secure resource sllocation for multiple-antenna NOMA with wireless power transfer. IEEE Trans. Green Commun. Netw. **2**(4), 1059–1071 (2018)

115. Cardellini, V., Valerio, V.D., Facchinei, F., Presti, F.L., Piccialli, V.: A game-theoretic approach to computation offloading in mobile cloud computing. Math Program. **157**(2), 421–449 (2015)

116. Jia, M., Cao, J., Liang, W.: Optimal cloudlet placement and user to cloudlet allocation in wireless metropolitan area networks. IEEE Trans. Serv. Comput. **5**(4), 725–737 (2017)

117. Lazar, A.: The throughput time delay function of an M/M/1 queue (Corresp.). IEEE Trans. Inf. Theory **29**(6), 914–918 (1983)

118. Deng, R., Lu, R., Lai, C., Luan, T.H., Liang, H.: Optimal workload allocation in fog-cloud computing towards balanced delay and power consumption. IEEE Internet Things J. **3**(6), 1171–1181 (2016)

119. Jiang, L., Tian, H., Xing, Z., Wang, K., Zhang, K., Maharjan, S., Gjessing, S., Zhang, Y.: Social-aware energy harvesting device-to-device communications in 5G networks. IEEE Wirel. Commun. **23**(4), 20–27 (2016)

120. Zhang, W., Wen, Y., Guan, K., Kilper, D., Luo, H., Wu, D.O.: Energy-optimal mobile cloud computing under stochastic wireless channel. IEEE Trans. Wirel. Commun. **12**(9), 4569–4581 (2013)

121. Liu, L., Chang, Z., Guo, X.: Socially aware dynamic computation offloading scheme for fog computing system with energy harvesting devices. IEEE Internet Things J. **5**(3), 1869–1879 (2018)
122. Machol, R.E.: Queue theory. IRE Trans. Educ. **E-5**(2), 99–105 (2007)
123. Xu, J., Hou, J., Tan, Y., Feng, E.: Exponential penalty function method for generalized nash equilibrium problem. Oper. Res. Manag. Sci. **24**(1), 81–89 (2015)
124. Xie, L.: The general convex smoothing problem solved by a semismooth newton algorithm. Math. Numerica Sinica. **27**(3), 257–266 (2005)
125. Sardellitti, S., Scutari, G., Barbarossa, S.: Joint optimization of radio and computational resources for multicell mobile-edge computing. IEEE Trans. Signal Inf. Process. Netw. **1**(2), 89–103 (2015)
126. Mao, Y., Zhang, J., Letaief, K.B.: Dynamic computation offloading for mobile-edge computing with energy harvesting devices. IEEE J. Sel. Area Commun. **34**(12), 3590–3605 (2016)
127. Chang, Z., Ristaniemi, T., Niu, Z.: Radio resource allocation for collaborative OFDMA relay networks with imperfect channel state information. IEEE Trans. Wirel. Commun. **13**(5), 2824–2835 (2014)
128. Hwang, D., Kim, D., Lee, T.: Throughput maximization for multiuser MIMO wireless powered communication networks. IEEE Trans. Veh. Technol. **65**(7), 5743–5748 (2016)
129. Yang, G. Ho, C.K., Guan, Y.L.: Dynamic resource allocation for multiple-antenna wireless power transfer. IEEE Trans. Signal Process. **62**(14), 3565–3577 (2014)
130. Son, H., Clerkx, B.: Joint beamforming design for multi-user wireless information and power transfer. IEEE Trans. Wirel. Commun. **13**(11), 6397–6409 (2014)
131. Hochwald, B., Marzetta, T., Tarokh, V.: Multiple antenna channel-hardening and its implications for rate feedback and scheduleing. IEEE Trans. Inf. Theory. **50**(9), 1893–1909 (2004)
132. Chang, Z., Wang, Z., Guo, X., Han, Z., Ristaniemi, T.:Energy-efficient resource allocation for wireless powered massive MIMO system with imperfect CSI. IEEE Trans. Green Commun. Netw. **1**(2), 121–130 (2017)
133. Chen, X., Wang, X., Chen, X.: Energy-efficient optimization for wireless information and power transfer in large-scale MIMO systems employing energy beamforming. IEEE Wireless Commun. Lett. **2**(6), 667–670 (2013)
134. Boyd, S., Vandenberghe, L.: Convex Optimization. Cambridge Univ Press, Cambridge U.K. (2004)
135. 3GPP TS 36.104: Evolved universal terrestrial radio access (E-UTRA), base station (BS) radio transmission and reception, Oct 2016
136. Hoglund, A., Van, D.P., Tirronen, T., Liberg, O., Sui, Y., Yavuz, E.A.: 3GPP Release 15 early data transmission. IEEE Commun. Standards Mag. **2**(2), 90–96 (2018)
137. 3GPP TS 36.300: Evolved universal terrestrial radio access (E-UTRA) and evolved universal terrestrial radio access network (E-UTRAN); Overall Description, Mar 2020

Printed in the United States
by Baker & Taylor Publisher Services